211
Advances in Polymer Science

Advances in Polymer Science
Recently Published and Forthcoming Volumes

New Frontiers in Polymer Synthesis
Volume Editor: Kobayashi, S.
Vol. 217, 2008

Polymers for Fuel Cells II
Volume Editor: Scherer, G. G.
Vol. 216, 2008

Polymers for Fuel Cells I
Volume Editor: Scherer, G. G.
Vol. 215, 2008

Photoresponsive Polymers II
Volume Editors: Lee, K. S., Marder, S.
Vol. 214, 2008

Photoresponsive Polymers I
Volume Editors: Lee, K. S., Marder, S.
Vol. 213, 2008

Polyfluorenes
Volume Editors: Scherf, U., Neher, D.
Vol. 212, 2008

Chromatography for Sustainable Polymeric Materials
Renewable, Degradable and Recyclable
Volume Editors: Albertsson, A.-C., Hakkarainen, M.
Vol. 211, 2008

Wax Crystal Control · Nanocomposites Stimuli-Responsive Polymers
Vol. 210, 2008

Functional Materials and Biomaterials
Vol. 209, 2007

Phase-Separated Interpenetrating Polymer Networks
Authors: Lipatov, Y. S., Alekseeva, T.
Vol. 208, 2007

Hydrogen Bonded Polymers
Volume Editor: Binder, W.
Vol. 207, 2007

Oligomers · Polymer Composites Molecular Imprinting
Vol. 206, 2007

Polysaccharides II
Volume Editor: Klemm, D.
Vol. 205, 2006

Neodymium Based Ziegler Catalysts – Fundamental Chemistry
Volume Editor: Nuyken, O.
Vol. 204, 2006

Polymers for Regenerative Medicine
Volume Editor: Werner, C.
Vol. 203, 2006

Peptide Hybrid Polymers
Volume Editors: Klok, H.-A., Schlaad, H.
Vol. 202, 2006

Supramolecular Polymers Polymeric Betains · Oligomers
Vol. 201, 2006

Ordered Polymeric Nanostructures at Surfaces
Volume Editor: Vancso, G. J., Reiter, G.
Vol. 200, 2006

Emissive Materials · Nanomaterials
Vol. 199, 2006

Surface-Initiated Polymerization II
Volume Editor: Jordan, R.
Vol. 198, 2006

Surface-Initiated Polymerization I
Volume Editor: Jordan, R.
Vol. 197, 2006

Chromatography for Sustainable Polymeric Materials

Renewable, Degradable and Recyclable

Volume Editors:
Ann-Christine Albertsson · Minna Hakkarainen

With contributions by
A.-C. Albertsson · L. Burman · M. Hakkarainen
M. Gröning · C. Strandberg

Springer

The series *Advances in Polymer Science* presents critical reviews of the present and future trends in polymer and biopolymer science including chemistry, physical chemistry, physics and material science. It is adressed to all scientists at universities and in industry who wish to keep abreast of advances in the topics covered.
As a rule, contributions are specially commissioned. The editors and publishers will, however, always be pleased to receive suggestions and supplementary information. Papers are accepted for *Advances in Polymer Science* in English.
In references *Advances in Polymer Science* is abbreviated *Adv Polym Sci* and is cited as a journal.

Springer WWW home page: springer.com
Visit the APS content at springerlink.com

ISBN 978-3-540-78762-4 e-ISBN 978-3-540-78763-1
DOI 10.1007/978-3-540-78763-1

Advances in Polymer Science ISSN 0065-3195

Library of Congress Control Number: 2008922725

Cover design: WMXDesign GmbH, Heidelberg
Typesetting and Production: le-tex publishing services oHG, Leipzig

Printed on acid-free paper

9 8 7 6 5 4 3 2 1 0

springer.com

Volume Editors

Ann-Christine Albertsson

Department of Fibre
and Polymer Technology
School of Chemical Science and Engineering
Royal Institute of Technology (KTH)
10044 Stockholm, Sweden
aila@polymer.kth.se

Minna Hakkarainen

Department of Fibre
and Polymer Technology
School of Chemical Science and Engineering
Royal Institute of Technology (KTH)
10044 Stockholm, Sweden
minna@polymer.kth.se

Editorial Board

Advances in Polymer Science Also Available Electronically

For all customers who have a standing order to Advances in Polymer Science, we offer the electronic version via SpringerLink free of charge. Please contact your librarian who can receive a password or free access to the full articles by registering at:

springerlink.com

If you do not have a subscription, you can still view the tables of contents of the volumes and the abstract of each article by going to the SpringerLink Homepage, clicking on "Browse by Online Libraries", then "Chemical Sciences", and finally choose Advances in Polymer Science.

You will find information about the

- Editorial Board
- Aims and Scope
- Instructions for Authors
- Sample Contribution

at springer.com using the search function.

Color figures are published in full color within the electronic version on SpringerLink.

Preface

Polymeric materials, both "inert" and degradable, constantly interact with the surroundings. Because of this interaction changes take place in the polymer matrix and small molecules are released to the environment. Reliable methods for testing biodegradability and environmental interaction of renewable resources and biodegradable polymers are required to answer the remaining questions concerning the environmental impact of these future materials. In the case of degradable polymers multiple factors affect the degradation process and small changes in the chemical structure or product formulation may change the susceptibility to degradation or cause different degradation product patterns, rendering the product less environmentally adaptable. Development of sustainable polymeric materials also demands the development of more migration-resistant polymer additives. Chromatographic techniques especially gas chromatography and liquid chromatography preferentially coupled to mass spectrometric detection are ideal tools for studying these low molecular weight compounds and polymer–environment interactions.

In the first chapter of this volume chromatographic fingerprinting and indicator product concepts are presented as tools for evaluating polymeric materials. These concepts have great potential in evaluation of degradation state and life-time/service-life of polymeric materials, evaluation of anti-oxidant or pro-oxidant systems, degradation mechanism and processing parameters as well as rapid comparison and quality control of materials. The solid-phase microextraction technique has rapidly found applications in numerous fields. The second chapter reviews the extraction of polymer degradation products and additives, monomer-rests, odour compounds, migrants from packaging and medical products as well as extraction of polymer additives from environmental samples and biological fluids by solid-phase microextraction demonstrating the high versatility and potential of this technique also in polymer analysis. In the third chapter the possibilities and limitations in the headspace extraction of volatiles from solid polymer matrixes are discussed. Examples of the use of multiple headspace extraction to remove matrix effects are shown and finally the application of headspace analysis for early degradation detection and quality control of recycled materials is presented. The fourth chapter summarises the literature on chromatographic analysis of degradation products from the most common aliphatic and aliphatic–aromatic polyesters. Especially the effect of macromolecular architecture and copolymer composition on the resulting degradation mechanism and degradation product pattern is discussed. The last two chapters deal with the analysis of polymer additives. The fifth chapter overviews different extraction techniques and aspects of analyzing antioxidants in polymeric materials. The sixth chapter discusses the migration of monomeric and polymeric PVC plasticizers with the focus on migration from medical products and food packaging. Especially the possibilities of improving the migration resistance and plasticizing properties of polymeric PVC plasticizers through the right plasticizer design are presented.

The interest in degradable and/or renewable materials is increasing rapidly. Degradation of these materials is still often studied only by measuring the weight loss or changes in molecular weight, which can be misleading. Especially in the case of bioresorbable materials the knowledge of degradation products is a crucial point for biocompatibility of the materials. As an example we have in chapter four presented results showing the influence of macromolecular design on the formation of acidic degradation products, a possible cause of negative impacts in the body. We have also shown that copolymer composition influences the stability, degradation mechanism and amount of degradation products formed during radiation sterilization. Hopefully these chapters will inspire more extensive use of chromatographic techniques for polymer analysis and result in increased understanding of polymeric materials, which in turn will provide tools for the development of sustainable future materials.

Stockholm, April 2008

Ann-Christine Albertsson
Minna Hakkarainen

Contents

Adv Polym Sci (2008) 211: 1–22
DOI 10.1007/12_2007_114

Published online: 21 April 2007

Indicator Products and Chromatographic Fingerprinting: New Tools for Degradation State and Lifetime Estimation

Lina Burman · Ann-Christine Albertsson (✉) · Minna Hakkarainen

Department of Fibre and Polymer Technology,
School of Chemical Science and Engineering,
Royal Institute of Technology (KTH), S-10044 Stockholm, Sweden
aila@polymer.kth.se

Abstract The demands on polymeric products are growing both with respect to their function and purity. There is a need for new high-throughput characterisation tools for rapid quality control and evaluation of materials. Precise control over degradation rate and service-life are also prerequisites for successful use of degradable polymers in an increasing number of applications. The chromatographic fingerprinting and indicator product concepts, presented in the current paper, are novel and attractive alternatives for rapid evaluation of the product quality, degradability, durability and service-life. The sensitivity of these techniques allows for detection of small initial changes in the materials and signs of early degradation. The possible applications include evaluation of different pro-oxidants or antioxidants, optimisation of processing parameters, evaluation of long-term properties or storage stability and lifetime prediction. The same principal could also be applied to process control and monitoring, acceptance or rejection of raw materials, intermediate and final products. The usefulness of indicator products and chromatographic fingerprinting is shown for estimation of the degradation state of degradable polyethylene. In addition, chromatographic fingerprinting together with multivariate data analysis is utilised to classify degradable polyethylene materials based on their incorporated pro-oxidant systems.

Keywords Chromatographic fingerprinting · Degradation · Indicator products · Lifetime prediction · Long-term properties

Abbreviations

ATD	automated thermal desorption
ATR-FTIR	attenuated total reflection-Fourier transform infrared spectroscopy
CL	chemiluminescence
CP	conducting polymers sensor
DSC	differential scanning calorimetry
FTIR	Fourier transform infrared spectroscopy
GC	gas chromatography
GC-MS	gas chromatography mass spectrometry
HDPE	high density polyethylene
LLDPE	linear low density polyethylene
LSE	liquid-solid extraction
MALDI	matrix-assisted laser desorption ionisation
MDA	multivariate data analysis
MFI	melt flow index
MOS	metal oxide semiconductors
MOSFET	metal oxide semiconductors field effect transistor
MS	mass spectrometry
NIR	near infrared reflection spectroscopy
PC	principal component
PCA	principal component analysis
PCL	polycaprolactone
PCR	principal component regression
PLLA	poly (L-lactic acid)
PLS	partial least squares regression
SEC	size exclusion chromatography
SPE	solid phase extraction
SPME	solid phase microextraction

1 Introduction

Throughout their life cycle, polyolefin's suffer oxidative degradation promoted by heat, UV-radiation and mechanical stress. The degradation is associated with irreversible changes in the chemical structure of the polymer. It influences the physical and chemical properties, such as morphology, molecular weight, tensile strength, elongation at break and colour. The new methods presented in this review for classification and rapid degradation state estimation are valuable tools for evaluation of polyolefin long-term properties and further for development of tailored polymer materials.

Degradable materials are desirable in various applications ranging from disposables, decreasing the amount of litter, to mulch films improving growth conditions for grain. Therefore, several different degradable polyethylene materials have been developed and are on the market today. The susceptibility of polyolefins to degradation can be varied by additives or by copolymerisation. Transition metal ions, e.g. iron, manganese and copper, catalyse the decomposition of hydro peroxides in thermal and photo-oxidation, see Eqs. 1 and 2 [1], and are used to enhance the degradation at low temperatures of otherwise relatively stable polymers such as polyethylene [2]. The products of the catalysed decomposition of hydro peroxides are similar to the products from un-catalysed oxidation processes [3].

Catalytic decomposition of hydro peroxides.

$$\mathrm{ROOH} + \mathrm{M}^{n} \rightarrow \mathrm{RO}^{\cdot} + \mathrm{OH}^{-} + \mathrm{M}^{n+1} \tag{1}$$

$$\mathrm{ROOH} + \mathrm{M}^{n+1} \rightarrow \mathrm{ROO}^{\cdot} + \mathrm{H}^{+} + \mathrm{M}^{n} \tag{2}$$

The use of transition metals as pro-oxidants in polyethylene gives degradable cost effective materials with good technical performance. Pro-oxidant systems may also contain natural polymers, such as starch, or unsaturated polymers [4–6]. The ability of materials to degrade by thermal oxidation and UV radiation as well as controlled degradation rate are crucial for their application. It is therefore important to investigate the degradation process during the early stages of oxidation to be able to produce materials for specific applications and for different degradation rates.

A good example of a class of materials with specific stability and degradability criteria are mulch films for corn production as seen in Fig. 1. They should protect the crops at the beginning of the season but be brittle enough

Fig. 1 Corn production with and without mulch-film

after 4 to 6 weeks for the crops to puncture the films without being damaged. However, the films must also be sufficiently resistant so that they are not torn into pieces by wind and normal weather conditions during the time when the plants are still small. With such demands it is crucial to understand and to have control over the degradation process during the early stages. To be able to detect small changes in the material would be very valuable for the development of rapid classification methods based on the initial stages of degradation. This would provide further tools for the development of improved degradable polyethylene materials and for making the right choices between the already existing ones.

Early degradation state detection is also a key issue in the field of stabilised materials [7]. The evaluation of long-term efficiency of antioxidants under non-accelerated conditions takes too much time to be practical. The accelerated tests currently in use are often made under unrealistic physical conditions leading to unreliable results [8–10]. Accelerated aging at high temperature is frequently used even though the degradation and stabilisation reactions taking place at high temperatures are different from those taking place at low temperatures. Sensitive techniques for early degradation detection are, thus, essential components in the effort to reduce the acceleration needed to reach practical test times.

2 Evaluation of Long-Term Performance of Polyethylene

Polymer degradation can be analysed at macroscopic, macromolecular or molecular scale, Fig. 2. The detection of early degradation and small differences in degradation behaviour between different materials requires an

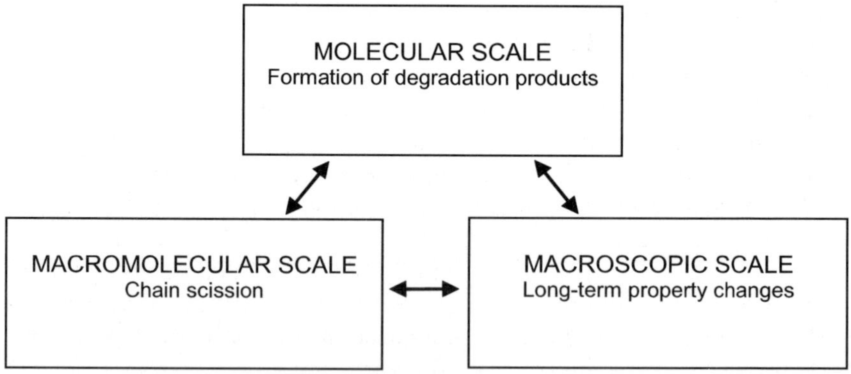

Fig. 2 Degradation of polymers can be analysed at macroscopic, macromolecular or molecular level

analytical technique with high sensitivity. Macroscopic scale properties are less sensitive parameters for early degradation detection and degradation state evaluations than the changes taking place at the molecular level, e.g., the formation of degradation products. The most common techniques used today to monitor the degradation of polymers are Fourier transform infrared spectroscopy (FTIR), where a carbonyl index is used as a measure of the oxidation induction time and degradation rate, size exclusion chromatography (SEC) to follow the changes in molecular weight [11, 12], differential scanning calorimetry (DSC) to follow the changes in crystallinity [13], and mechanical testing of changes in strength and brittleness. As an example loss in mechanical properties occurs first when the molecular weight of the polymer has decreased to a critical value [14].

2.1 Evaluation Based on Changes at the Molecular Level

Degradation products can usually be detected, identified and quantified long before the mechanical performance of the material changes. Degradation products also give information regarding the degradation mechanisms beyond these changes. FTIR is a useful technique that provides information at a molecular level but compared to chromatographic techniques there are limitations in how detailed the information is that is obtained. Carbonyl compounds account for most of the oxidation products and they are seen in the FTIR spectra in the region between 1680 and 1780 cm^{-1} as overlapping bands corresponding mainly to acids (1712 cm^{-1}), ketones (1720 cm^{-1}), aldehydes (1730 cm^{-1}), esters (1743 cm^{-1}) and lactones (1785 cm^{-1}) [15, 16]. Because of the overlapping of the bands, derivatisation, using for example NO and SF_4, is necessary for quantification of the functional groups. [17]. During the 1970s and 1980s Albertsson et al. followed the degradation of polyethylene by measuring the CO_2 emission from the polymers using a ^{14}C technique with liquid scintillation spectrometry [18–20]. The labelling assured that the CO_2 containing ^{14}C came from degrading polymers. Chemiluminescence (CL) is a newer technique that has mostly been used for evaluation of stabiliser efficiency, but that today is sensitive enough for early degradation detection and classification of degradable polyethylene materials [21]. The counted photons emitted from the oxidising polymer correlate with the amount of hydro peroxides in the material, i.e. the initial degradation products during oxidation [22, 23]. CL has been shown to detect oxidation earlier than FTIR [21] in degradable polyethylene, even when looking at the range of the hydro peroxide detection in the FTIR spectra. However, good degradation state estimations are prevented by a non-linear increase in the luminescence intensity versus the degradation time.

2.1.1
Evaluation Using Gas Chromatography

FTIR is a valuable technique for obtaining information of the various product groups. However, identification of the individual degradation products gives more insight into the degradation mechanisms [24]. Gas chromatography (GC) in combination with selective extraction method and mass spectrometric detection are the ideal tools for identification of volatile and semi-volatile products. Since the early 1980's Albertsson et al. have used chromatographic analyses of low molecular weight degradation products to study the long-term performance of polymers [25, 26]. Their latest works on degradable polyethylene have focused on the development of rapid and informative tools to provide a greater understanding within this area. Reliable extraction methods are vital for the correct chromatographic analysis of long-term performance of polymers. The choice of extraction technique depends on the analytes, on the surrounding media and on the purpose of the extraction. The development of several extraction methods utilising liquid-solid extraction (LSE), solid phase extraction (SPE) and solid phase microextraction (SPME), made it possible to identify over 200 degradation products and product fingerprints to clarify the complex degradation patterns of polyethylene [27–32]. The most abundant groups of degradation products were mono- and dicarboxylic acids, but alkanes, alkenes, ketones, aldehydes and alcohols were formed as well.

2.1.2
Multivariate Data Analysis for Optimised Information Extraction

The amount of data that can be obtained from, for example chromatographic and spectrometric techniques, has increased dramatically. Wold [33] introduced the principle of multivariate data analysis (MDA) in the mid-1970s as a way to obtain as much information as possible from these analyses. Estimates based on many variables have in addition the advantage of being more robust than estimates from a few measurements since the first are decided with higher degrees of freedom [34].

Principal component analysis (PCA) is a qualitative method where the X-data can be studied without any knowledge of the Y-data. A score plot of the X-data gives an overview of possible patterns in the data and is therefore a useful tool for classification. The X-data are explained using uncorrelated vectors in pairs called principal components (PC). The first principal component (PC1) is in the direction of the largest variation in the multidimensional X space, Fig. 3. PC2 is in the direction of the second largest variation perpendicular to PC1 etc, all orthogonal to each other. The two-dimensional plane containing two principal components, e.g. PC1 and PC2, is called a score plot. The number of components to be included in the model is chosen on the basis of the amount of variation in the data that each of

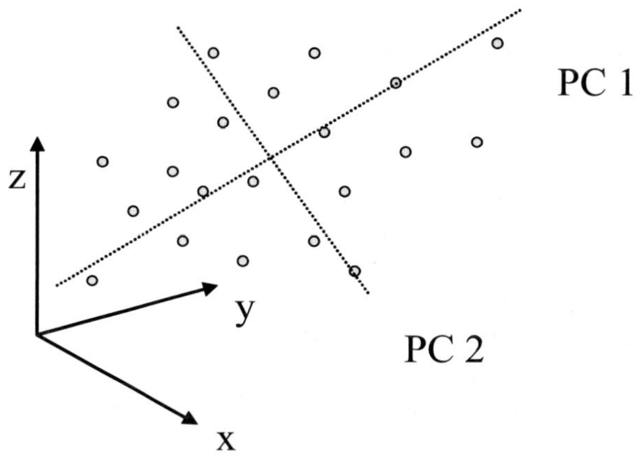

Fig. 3 Position of principal components in a PCA model

the components describes. The number of components shall be sufficient to explain all the systematic deviations. The residual difference between the number of components and the number of variables in the data matrix correspond to noise and is thus excluded.

The prediction of Y-data of unknown samples is based on a regression method where the X-data are correlated to the Y-data. The multivariate methods, usually used for such a calibration, are principal component regression (PCR) and partial least squares regression (PLS). Both methods are based on the assumption of linearity and can deal with co-linear data. The problem of co-linearity is solved in the same way as the formation of a PCA plot. The X-variables are added together into latent variables, score vectors. These vectors are independent since they are orthogonal to each other and they can therefore be used to create a calibration model.

In PLS, the part of the X-data that best describes the Y-data is searched for. This is the direction in the variable space in which the co-variation between the X and the Y variables is greatest. The score vectors are here described with the help of a unit vector w that weights the columns in the X-data matrix according to how well they describe variations in Y [35]. There are two kinds of PLS, viz. PLS 1 and PLS 2 [36]. PLS 1 can only be used for calibration towards one Y-variable at a time whereas PLS 2 can be used for the calibration of many Y-variables simultaneously in the same model. PLS 2 gives a more general model with a poorer prediction especially in the case of non-linearity, but it can be better than PLS 1 if there is a co variation between the samples. A PLS model can deal only with values that fit into its domain. The model does not take into consideration the possibility that the calibration curve can be non-linear at high and low concentrations.

In the case of multivariate calibration, the variables can have different units. It is then necessary to scale these variables to eliminate differences in their relative sizes so that they are all taken into consideration. The variables are scaled by deviating each of them with its standard deviation. The variables are usually also centered to make sure that the largest variation in the space is described by the principal components [37].

As an example, multivariate data analysis has been used to correlate spectral responses of fluorescence with different stabilisers for evaluation of their process stabilisation efficiency in polypropylene [38]. PLS analyses were then performed on the variation in the spectra as a function of the melt flow index (MFI), which is a commonly used parameter for ranking of stabilisers. The predicted MFI values correlated well with the measured values. Multivariate data analysis has also been used to evaluate and predict the degradation rate of glass fibre reinforced polyester composites by comparing the degradation product patterns with aging time and temperature [39]. In a recent study the effect of polyester molecular weight, end-groups and chain architecture on migration rate of poly(vinyl chloride)/polyester blends was evaluated to develop a migration resistant polymeric plasticiser [40].

3 Indicator Products

The demands on polymeric products are growing both with respect to their function and purity. There is a need for new high-throughput characterisation tools for rapid quality control and evaluation of materials. If there is a connection between the molecular, macromolecular and macroscopic scale changes in polymeric materials, then macromolecular and macroscopic scale changes could be predicted from the molecular scale events. Analysis of certain volatiles or chemical markers is widely used to rapidly and reliably evaluate the quality and shelf-life of food. The condition of different plants can also be deduced from the volatiles they are emitting. This has encouraged us to find corresponding methods for evaluation of polymeric materials, where the formation of certain volatiles or indicator products is related to the condition of the material.

Electronic nose technology and analysis of volatiles has long been applied in the food industry to control the quality of food products and to determine shelf lives. For example, sensor arrays based on different SnO_2 gas sensors can be used to distinguish milk products of different rancidity levels [41]. Standard microbial test prediction of shelf life of milk products has a low level of reliability due to relatively poor correlation between microbial counts and actual shelf life. Several alternative methods have therefore been developed. One method is based on dynamic headspace capillary gas chromatography analyses of volatiles in milk followed by MDA analyses. [42]. Principals of this method were later used for development of a faster and simpler test, where the extraction was performed by the SPME technique, the extracts were analysed by GC-MS, which was followed by MDA analysis based on PLS modelling [43]. In addition, this method was more effective in extraction of volatile fatty acids, an important group of contributors to off-flavour and odour in milk. The electronic nose technology has also been proven useful for quality evaluation of fish and fruits [44, 45].

In the area of polymers, electronic nose technology has been evaluated in a couple of papers during the last years for quality control in the automotive industry [46–49]. The evaluations have included several different kinds of sensors such as metal oxide semiconductors (MOS), metal oxide semiconductor field effect transistor (MOSFET), conducting polymer sensors (CP) and MS devices. The odour in new cars is caused by a large number of different compounds. MOS was shown to be the most effective technique at a comparison between the effectiveness of several MOS, MOSFET and CP instruments in discriminating polyurethane foams from each other in the interiors of cars [47]. The next best was MOSFET in combination with MOS while the CP instrument was non-discriminating. All of these gas sensors are based on a variation in the electrical resistance when volatile components are present. MDA was used in all the cases for the analyses. The separation between the groups in the PCA score plots depended a lot on the number of sensors used in the individual instruments. By comparing GC and GC-MS data and sensory evaluation by a test panel, the identified and quantified compounds could be correlated to different types of odours, e.g. unpleasant and stimulating odours [46]. In this case 20 words were used to classify the type of odour with a PCA model. The detection limit was taken into consideration at estimation of the present compounds contribution to the odour. An alternative to GC-MS is using only a MS device [48]. With MS-based chemical sensors, the sample in gas form is introduced directly to the MS analysator without a pre-separation step, leading to overlapping spectra. Ion fragments in the spectra represent the sensors.

Metal oxide semiconductor chemical sensors in combination with MDA have been shown to be useful to estimate the oxidative stability of polypropylene during processing instead of traditional melt flow index analysis [50]. An array of sensors was used to receive a detailed analysis of volatiles. At quality measurements of different poly(butylene adipate)s the use of indicator products has been proven better than analyses of the decrease in molecular weight or mass loss for early degradation detection. Adipic acid, quantified using gas chromatography, was then used as the indicator product [51].

3.1 Degradation State Estimation

If a correlation can be established between the matrix changes and formation of certain degradation products, then these compounds can be used as indicator products to estimate the degree of degradation and the remaining lifetime of the polymer product. Hakkarainen et al. have investigated this possibility for several polyethylenes and found butanedioic acid and butyrolactone to be suitable indicator products for predicting oxidative degradation of polyethylene [52]. The changes in the amount of degradation products were compared with changes in molar mass and the number of chain scissions. Starch/polyethylene and starch/polyethylene vinyl acetate were thermo-oxidised, at 190 °C or 230 °C for 30 min or3 h. In addition, several polyethylenes containing different photo initiators and stabilisers were photo-oxidised for 300 h. The key questions to answer were: what compounds are released during thermo- and photo-oxidation and which of these compounds are the best predictors of matrix deterioration. After oxidation, degradation products of the groups alkanes, alkenes, alcohols, ketones, aldehydes, carboxylic acids, dicarboxylic acids and lactones were extracted from the degraded materials.

During thermo-oxidation, especially the amount of dicarboxylic acids and lactones increased as the molar mass decreased. Further studies were therefore performed with the focus on these groups of degradation products to investigate if the formed amounts could be correlated to the change in molar mass. The samples in Fig. 4 are arranged as a function of decreasing number average molar mass or increasing degree of degradation. It is clearly seen that there is a good correlation between the relative amount of dicarboxylic acids and lactones extracted from the samples and the changes in their number average molar masses after aging at the two different temperatures. The total amount of indicator products increased irrespective of the aging temperature as the molar mass decreased. However, the relative amount of lactones in relation to the relative amount of dicarboxylic acids was higher in the samples aged at the higher temperature.

Predicting the degree of degradation would be further simplified if only one or two compounds could be used as indicator products instead of whole classes of compounds. Butanedioic acid was the most abundant, or one of the most abundant, dicarboxylic acids in the degradation product patterns and was therefore chosen as a suitable candidate to represent the group of dicarboxylic acids. A comparison between the relative amount of butanedioic acid to the relative amount of all the dicarboxylic acids showed that the relative amounts correlated quite well. Similarly, the amount of butyrolactone, by far the most abundant of the lactones, correlated well with the total amount of lactones.

The relative amount of indicator products formed during oxidation was also plotted versus the number of chain scissions to investigate if they correlated. Three different cases were studied, i.e. all dicarboxylic acids and lactones, all dicarboxylic acids or only butanedioic acid as an indicator product. During oxidation at 190 °C the best correlation was obtained when only butanedioic acid was used as an indicator product, Fig. 5. However, it was concluded that if different aging temperatures are used, then both dicarboxylic acids and lactones have to be used as indicator products.

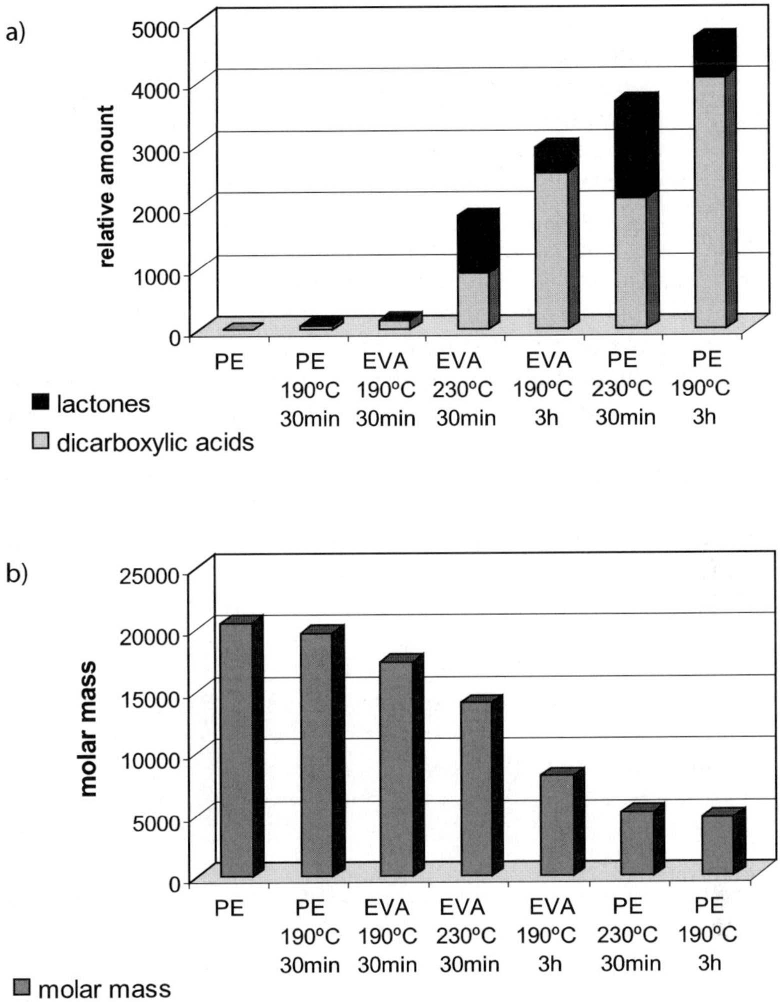

Fig. 4 **a** Relative amount of indicator products, i.e. dicarboxylic acids and lactones formed during thermo-oxidation of PE and EVA at 190 and 230 °C. **b** Number average molar mass after thermo-oxidation. Reprinted with permission from [52]

During photo-oxidation, dicarboxylic acids were the class of products that clearly increased in the most severely degraded samples. As during thermo-oxidation, the most abundant of the dicarboxylic acids was butanedioic acid. Comparison between the number average molar mass and the relative amount of butanedioic acid, Fig. 6, showed a connection between the formation of butanedioic acid and the degree of degradation in the polyethylene matrix. However, the relative sum of all the carboxylic acids correlated even better with the number of chain scissions than the amount of only butanedioic acid, Fig. 7.

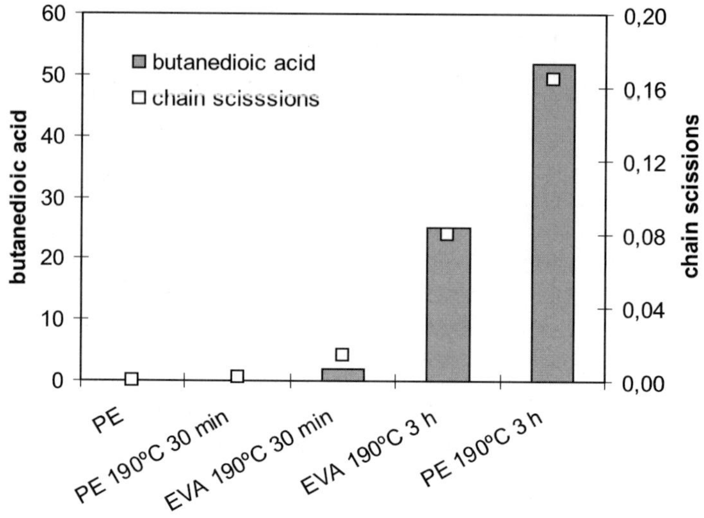

Fig. 5 Correlation between the number of chain scissions and the formation of butanedioic acid during aging at 190 °C

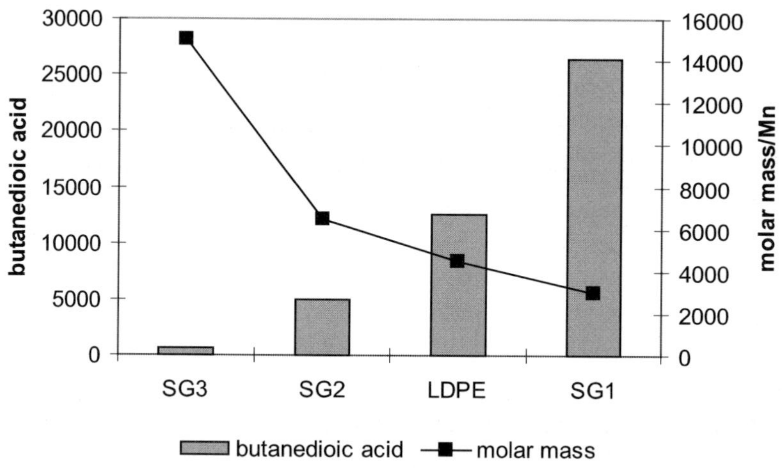

Fig. 6 Comparison between the number average molar mass and the relative amount of butanedioic acid in photo-oxidised polyethylenes containing different photoinitiators and stabilisers. The used additives were iron dimethyldithiocarbamate in SG1, iron dimethyldithiocarbamate and 0.8% carbon black in SG2 and iron dimethyldithiocarbamate and nickel dibutyldithiocarbamate in SG3

A correlation between matrix changes and formation of certain degradation products has also been found for polyamide 6.6 [53]. Especially the amount of cyclic imides increased during thermo-oxidation, of which 1-pentyl-2,5-pyrrolidinedione was the most abundant. The release of this

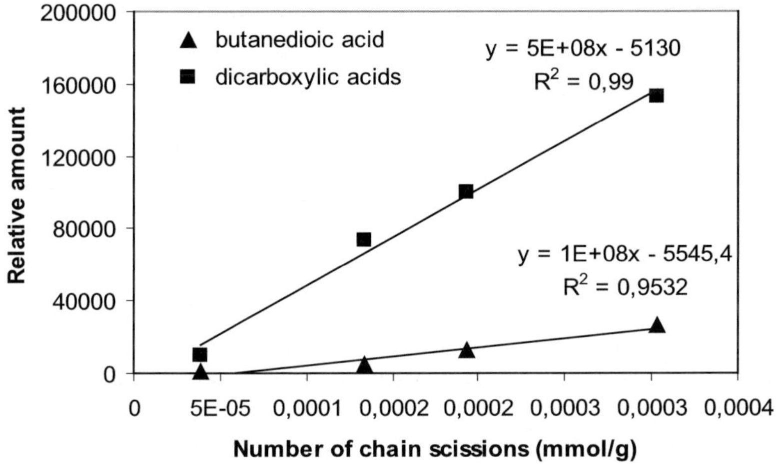

Fig. 7 The relative amount of indicator products versus the number of chain scissions for photo-oxidised polyethylenes containing different photoinitiators and stabilisers. The used additives were iron dimethyldithiocarbamate in SG1, iron dimethyldithiocarbamate and 0.8% carbon black in SG2 and iron dimethyldithiocarbamate and nickel dibutyldithiocarbamate in SG3

cyclic amide correlated well with changes in tensile strength for both virgin and recycled polyamide. A linear relationship has also been found between the number of chain scissions and weight loss, i.e. total amount of volatile degradation products formed during thermal degradation of poly(2-hexane) [54].

4 Chromatographic Fingerprinting

Chromatographic analysis of the degradation products after thermal oxidation and UV radiation of polymers gives rise to complex chromatographic fingerprints [28, 32, 55]. These fingerprints can be used in many applications, such as quality control of synthesis, environmental interactions, degradation mechanism evaluations and stability predictions. The pattern of low molecular weight molecules emitted from a polymeric material can for example tell you whether the material is good enough for recycling [53].

Chromatographic fingerprinting has been shown to be a useful technique to differentiate between abiotic and biotic degradation and for evaluation of the environmental impact of degradable polymers [56, 57]. A homologous series of mono- and dicarboxylic acids and ketoacids were extracted from degradable polyethylene aged at ambient temperature in an abiotic environment after pre-oxidation. However, the acids completely disappeared during aging in biotic mineral medium (*Arthrobacter paraffineus*) indicating assimilation of the acids by micro-organisms. The important question concerning the environmental impact of degradable polymers "Are the low molar mass degradation products assimilated by micro-organisms or do they accumulate in the environment?" could thereby be answered by looking at the differences in the product patterns after aging in a biotic and abiotic environment.

Similar studies have also been performed with poly (L-lactide) (PLLA) and polycaprolactone (PCL) [58–60]. The PLLA films originally contained several low molecular weight products [58]. Lactide and lactoyl lactic acid, the acyclic dimer, were the major products in the chromatograms followed by lactic acid. In the biotic environment lactic acid and the lactoyl lactic acid, were both rapidly assimilated by the micro-organisms. However, in the abiotic environment increasing amounts of lactic acid and lactoyl lactic acid were detected during the whole hydrolysis period. During the first weeks acetic acid and propanoic acid could also be extracted from the biotic mineral medium [59]. These are well-known fermentation products of lactic acid. As for the PLLA, several low molecular weight products were already present in the unaged PCL films [60]. The amount of cyclic dimer and trimer remained rather constant during abiotic ageing, but they were no longer detected after 2 weeks in the biotic medium due to assimilation by the micro-organisms.

Willoughby et al. have shown that by using automated thermal desorption-GC/MS (ATD-GC/MS) it is possible to fingerprint low molecular weight molecules entrapped in polymers to distinguish between stereo regular polyethylenes, and even powdered and pelletised grade of the same polymer [24]. The identified products were mostly hydrocarbons. The analysed volatile hydrocarbons from LLDPE could be correlated to the polymers base of C2 olefins as they all had an even number of carbons. Further, a higher proportion of unsaturated hydrocarbons were detected in HDPE than in LLDPE. A powdered and pelletised form of the same polymer could be distinguished both from total ion output and selective ion output. The differences in the fingerprints were derived from easier evaporation from the powdered form, but also from differences in their thermal history. Analyses of the thermally treated pellets showed different relative abundance of the ion fragments in the MS and lower retention times in the total ion chromatogram than the powder. This was explained as an effect of skeletal rearrangements and inter conversion between alkane and alkene. They found that the m/z ratio 57/43 had the potential to discriminate between powder and pellets and the m/z ratio 41/(43+57) could discriminate between LLDPE and HDPE. Differences were also seen between powdered and pelletised polypropylene.

Matrix assisted laser desorption ionisation (MALDI) is an important technique for characterisation and fingerprinting of large molecules. However, solvents used at sample preparation have a negative effect on the quality of the fingerprint. Work is therefore performed on development of solvent free sample preparation methods for characterisation of synthetic polymers [61].

When the matrix and oligomers are co-crystallised from the condensed phase, problems arise with discrimination, decreased sensitivity, selectivity and reproducibility. Avoiding solvents is a way to avoid effects of, for example, polymer solubility and pH.

Fingerprinting has also been shown useful in evaluation of spectrometric data. In an evaluation of the composition of copolymers, fingerprinting from attenuated total reflection-FTIR (ATR-FTIR) spectra was used followed by MDA [62]. The copolymerisation was monitored on-line. Plots of the covariance of the absorbance clearly showed which bands of wavelength correlate with the ethene and propene contents in the copolymers. The technique was also shown to be useful for copolymers of ethene and 1-hexene giving information about catalyst activity, polymerisation kinetics and the degree of homogeneity in the final copolymer among others [63]. Similar evaluations were later successively performed using spectra from near infrared reflection spectroscopy (NIR) followed by MDA for the copolymers [64].

4.1
Classification and Degradation State Estimation

The use of chromatographic fingerprinting for classification and degradation state estimation has been investigated by Albertsson et. al. [65]. The major hypothesis was that different pro-oxidant systems will lead to different degradation mechanisms and this would consequently affect the chromatographic fingerprints. Chromatographic fingerprinting of the product patterns enabled classification according to the type of pro-oxidant system, and degradation state estimation; i.e. the materials state in relation to its own lifetime cycle. This gives valuable possibilities for product control before use and for degradation state estimations of materials in use. For the investigation, chromatographic fingerprinting with the focus on the carboxylic acids, the most abundant group of degradation products in an abiotic environment, [66–68] was used in combination with multivariate data analysis. The evaluations were performed with four different polyethylene films thermally oxidised at 80 °C. One of the films was a reference material without pro-oxidants, two contained metal stearate pro-oxidant systems and one contained a pro-oxidant system with both a metal stearate and a polymer containing double bonds, viz. polyoctylene. The carboxylic acids in the oxidised samples were extracted with organic solvent, followed by analysis using GC-MS for relative quantification.

Chromatographic fingerprinting demands comparable patterns, which does not necessarily mean that complicated extraction procedures are necessary as long as the products of interest can be relatively quantified with the aid of a suitable internal standard. Low molecular weight carboxylic acids in the polyethylene films were extracted and simultaneously methylated using acidified methanol to decrease the acids polarity and thereby increasing the volatility of the acids [29,66]. Thereafter, the solvent was changed, by evaporation, to n-hexane to achieve better chromatograms and to detect a larger number of carboxylic acids than a direct analysis of the acidified methanol would enable. The simultaneous evaporation of low molecular weight acids during the change of solvent was compensated for by calculation of the individual degrees of evaporation of the acids. Chromatograms of methanol and hexane extracts can be seen in Fig. 8. The responses of the acid peaks relative to one of the internal standards, 5-phenylvaleric acid, and three large acids, methylated octadecanoic acid (m18), methylated heptadecanedioic acid (di17) and octadecanedioic acid (di18) in the methanol extract were compared with the corresponding values in the final extract (n-hexane), and this gave the percentage decrease. Although additional evaporation during the

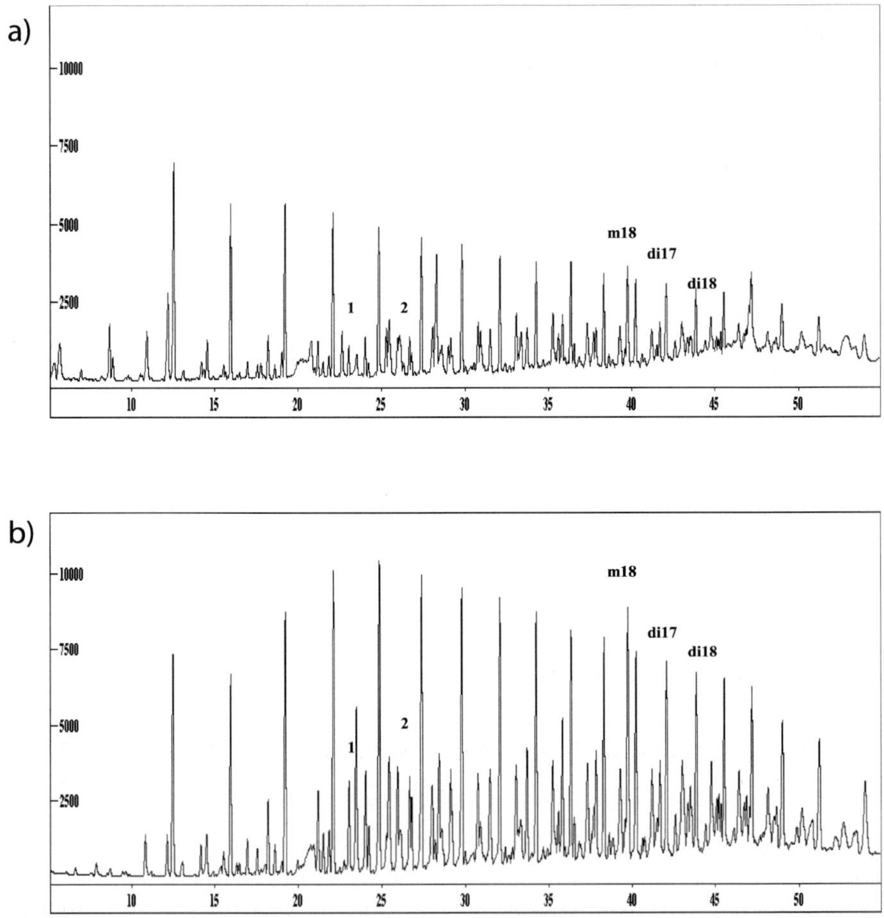

Fig. 8 Comparison between **a** extraction using acidified methanol and **b** solvent exchange to hexane

degradation, methylation and analysis steps occurs and the molecular-size-dependence of the responses affects the response values for the methylated acids, the samples could still be compared.

The chromatographic fingerprint of the dicarboxylic acids from the films containing polyoctylene clearly differed from the others. Diacids with five to nine carbons dominated after 4 days of oxidation and continued to dominate throughout the whole testing period. Fourier transform infrared spectroscopy (FTIR) was useful as a complement to the GC-MS analyses since it has the advantage that the functional groups in the samples can be detected independently of the volatility of the molecules. According to the FTIR spectra in Fig. 9 of samples containing polyoctylene, unaged and aged for 4 days, the vinyl groups in the polyoctylene (980 cm^{-1} [69]) were involved in reactions in the initial stage of the degradation. Possible reaction mechanisms leading to the simultaneous disappearance of the double bonds and formation of diacids with five to nine carbons, without traces of the intermediate formation of monoacids, were suggested to be zip depolymerisation of the polyoctylene through a cyclic transition state [32], Hock cleavage of allylic hydro peroxides [70–72], and to some extent formation of networks, which would saturate the double bonds.

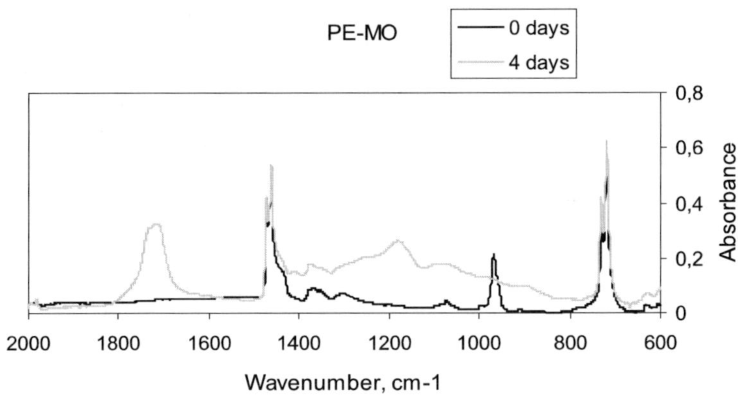

Fig. 9 FTIR spectra for polyethylene films containing 10% Fe (II)-stearate MB and 10% polyoctylene (PE-MO), unaged and aged for 4 days at 80 °C

In a PCA model over dicarboxylic acids from degradation of the degradable polyethylenes, Fig. 10, PC1 was totally settled by the time factor, whereas the class separation was described in PC2. The degraded samples were positioned at different sides of the mid-axes depending on the type of pro-oxidant system.

The position of the samples along PC1 in the PCA model indicated that it should be possible to determine the degree of degradation from the carboxylic acid patterns using PLS modelling. The grouping of the pro-oxidants indicated, however, that this is only possible for the polyethylene films that contain the same type of pro-oxidant system. From a scatter plot of a model based on one of the materials containing metal stearate pro-oxidant with predicted values of the other two materials containing pro-oxidants included, it was obvious that separate models have to be used for materials with different types of pro-oxidant systems. Most of the samples containing both metal stearate and a polymer in its pro-oxidant system were far from the ellipse that corresponds to the region in which samples are well described by the model. The obtained models show the methods potential for classification and degradation state estimation, but to obtain a good prediction model polyethylene films with different classes of pro-oxidants need to be analysed.

In the case of the monoacids, PC1 and PC2 were directed by the extracted hexadecanoic acid and octadecanoic acid originating from the stearate included in the pro-oxidant systems. The influence of the initial conditions on the X-data for the monoacids gave an opposite positioning of the results in the score plot of the PCA model. This was seen as a relatively narrow region, close to the central horizontal axis containing all of the results for the 24-day aged samples. The remaining samples diverged along the first PC to the right-hand side of the ellipse, with the least degraded ones at the far right-hand side and a separation along PC2 based on differences in the composition of the added stearate rather than the type of pro-oxidant system. The closely located pos-

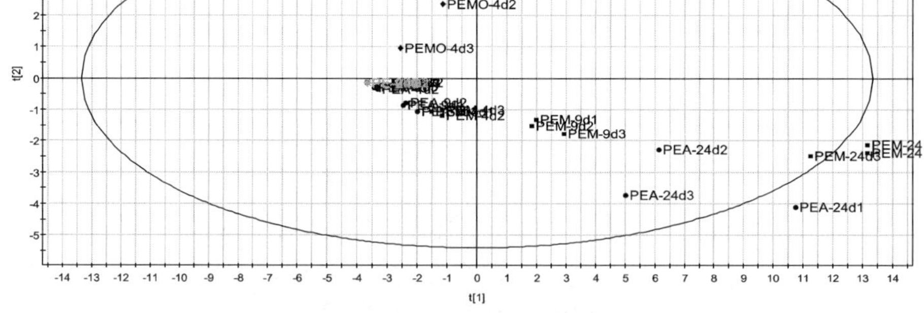

Fig. 10 Score scatter plot of the first two principal components from the PCA model for the diacids extracted from polyethylene films containing 10% Fe (II)-stearate MB (PE-M), class 1, 7.5% Ampacet MB (PE-A), class 2, 10% Fe (II)-stearate MB and 10% polyoctylene (PE-MO), class 3, and pure LLDPE (PE), class 4, unaged and aged for 4, 9 and 24 days at 80 °C. Reprinted with permission from Elsevier [65]

itions of the results for the most degraded samples and the relative positions of the results along the first PC indicate that monoacids cannot be used to estimate degradation states, and that they would negatively affect a degradation state model including all the acids.

Unpublished data of chromatographic fingerprints of diacids after photo-oxidation of the films revealed during the initial stages of degradation, similar patterns as for the thermally degraded films, Fig. 11. The materials were degraded using an accelerated weathering tester in cycles with UV radiation and condensation. The samples were therefore exposed mainly to photo oxidation, but also to some extent thermo oxidation and hydrolysis. The equipment was a QUV/SE with eight UVA-340 lamps that were programmed to give cycles with 8 h condensation at 50 °C and 16 h UV at 60 °C and 0.77 W/m^2/nm continuously during the aging. The carboxylic acids were extracted from the degraded materials using acidified methanol. A lower number of high molecular acids were therefore extracted. In addition propanedioic acid, not monitored after thermo-oxidation, was detected. The change of extraction solvent to hexane, as was done for the thermally aged materials, was not performed since the low molecular weight acids appeared to have a larger influence on the classification of the materials than the high molecular weight acids. The column used for the GC-MS analyses was a CP wax 58 from Varian. The differences in the diacid patterns seen in Fig. 11 are much smaller than in the case of the thermally oxidised samples. However, it shows the potential of classification possibilities of degradable polyethylenes from their combined photo and thermal oxidation durability, if tests are performed on

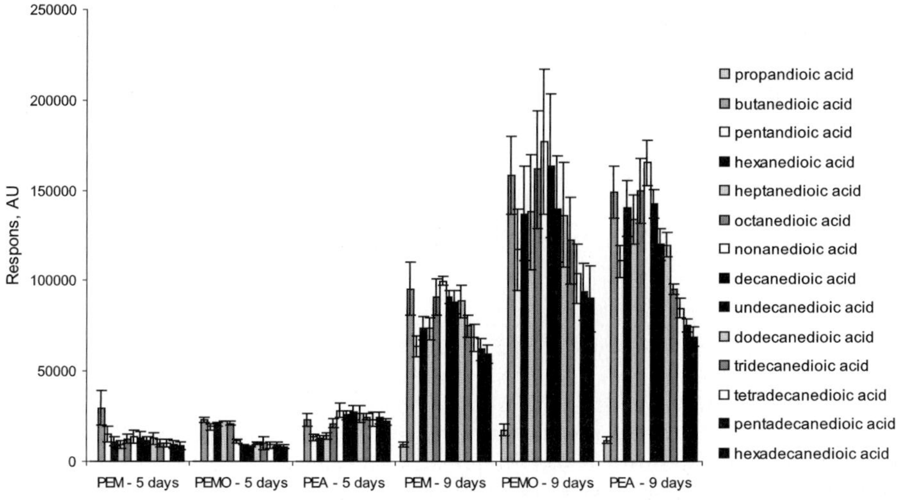

Fig. 11 Average detection responses of dicarboxylic acids extracted from polyethylene films containing 10% Fe (II)-stearate MB and 10% polyoctylene (PE-MO), 10% Fe (II)-stearate MB (PE-M), and 7.5% Ampacet MB (PE-A) photo oxidised for 5 and 9 days

samples from the early stage of the degradation. The differences in the degradation product patterns decreased at prolonged degradation since the double bonds in the pro-oxidant system react in the early stage of the degradation. Therefore, the severely degraded samples, in this case the samples degraded for 9 days, would highly affect the group formations in a PCA model where patterns from these samples are included.

5 Concluding Remarks

Early degradation detection requires sensitive analysis techniques. The first changes in the material will occur at a molecular scale. Chromatographic techniques are outstanding for such analyses. The high sensitivity of chromatographic techniques together with the large amount of data gives a good base for both early degradation detection and degradation state and life time estimation. The methods based on indicator products and chromatographic fingerprinting presented in the current paper are two examples where specific data is selected from a complex matrix of data obtained by chromatographic analyses. By careful choice of information from this data, valuable correlations with macromolecular scale changes were found. By using indicator products, a simpler method can be obtained compared to fingerprinting since only a few degradation products and their amounts have to be taken into consideration instead of total patterns of products. After finding suitable indicator products the method can be further simplified by choosing more selective extraction and analysis methods with a focus on the specific products. Chromatographic fingerprinting takes more information into consideration, especially in combination with multivariate data analysis. It can then simultaneously give information, for example, evaluation of degradation mechanisms and classification of degradable polyethylene materials on the basis of their incorporated pro-oxidant systems.

References

1. Allen NS (1992) Polymer Structure and Stability. In: Allen NS, Edge M (eds) Fundamentals of polymer degradation and stabilisation. Elsevier Applied Science, London, p 1
2. Albertsson A-C (1978) J Appl Polym Sci 22:3419
3. Scott G (1993) Initiators, prooxidants and sensitizers. In: Scott G (ed) Atmos Oxid Antioxid. Elsevier, Amsterdam, p 83
4. Griffin GJL (1977) US Patent 4 016 117
5. Scott G (1978) US Patent 4 121 025
6. Griffin GJL (1988) UK Patent PCT/GB88/00386
7. Burman L, Albertsson A-C (2005) J Polym Sci A Polym Chem 43:4537

8. Billingham NC, Calvert PD (1980) Dev Polym Stab 3:139
9. Allen NS, Edge M (1992) Thermal Stabilisation of Polymers. In: Allen NS, Edge M (eds) Fundamentals of Polymer Degradation and Stabilisation. Elsevier Applied Science, London, p 22
10. Pospisil J, Horak Z, Pilar J, Billingham NC, Zweifel H, Nespurek S (2003) Polym Degrad Stab 82:145
11. Al-Malaika S, Chohan S, Coker M, Scott G, Arnaud R, Dabin P, Fauve A, Lemaire JM (1995) J Macromol Sci Pure Appl Chem A32:709
12. Erlandsson B, Karlsson S, Albertsson A-C (1997) Polym Degrad Stab 55:237
13. Billingham NC, Prentice P, Walker TJ (1977) J Polym Sci Polym Symp 57:287
14. Audouin L, Langlois V, Verdu J (2004) J Mater Sci 29:569
15. Carlsson DJ, Wiles DM (1969) Macromolecules 2:587
16. Adams JH (1970) J Polym Sci Polym Chem Ed 8:1077
17. Lacoste J, Carlsson DJ (1992) J Polym Sci A Polym Chem 30:493
18. Albertsson A-C, Rånby B (1976) Biodegradation of synthetic polymers: the C-14 method applied to polyethylenes. In: Sharpley JM, Kaplan AM (eds) Proc 3rd Int Biodeg Symp. Applied Science Publishers, London, p 743
19. Albertsson A-C (1977) Thesis, Royal Institute of Technology, Stockholm
20. Albertsson A-C, Karlsson S (1988) J Appl Polym Sci 35:1289
21. Ragnarsson L, Albertsson A-C (2003) Biomacromolecules 4:900
22. Billingham NC, Then ETH, Gijsman PJ (1991) Polym Degrad Stab 34:263
23. Ahlblad G, Reitberger T, Terselius B, Stenberg B (1999) Polym Degrad Stab 65:179
24. Willoughby BG, Goldby A, Daviess J, Cain R (2003) Polym Test 22:553
25. Albertsson A-C, Banhidi ZG (1980) J Appl Polym Sci 25:1655
26. Karlsson S, Banhidi E, Banhidi ZG, Albertsson A-C (1984) Accumulation of Malodorous Amines and Polyamines Due to Clostridial Putrefaction. In: Berglund B, Lindberg T, Sundell J (eds) Indoor Air 3: Sensoring and Hyperactivity Reactions to Sick Buildings. Liber Tryck AB, Stockholm, p 287
27. Albertsson A-C, Barenstedt C, Karlsson S (1992) Polym Degrad Stab 37:163
28. Hakkarainen M, Albertsson A-C, Karlsson S (1996) J Chromatogr A 741:251
29. Khabbaz F, Albertsson A-C (2000) Biomacromolecules 1:665
30. Albertsson A-C, Andersson SO, Karlsson S (1987) Polym Degrad Stab 18:73
31. Albertsson A-C, Barenstedt C, Karlsson S (1994) J Appl Polym Sci 51:1097
32. Karlsson S, Hakkarainen M, Albertsson A-C (1997) Macromolecules 30:7721
33. Wold S (1976) Pattern Recogn 8:127
34. Bro R (2003) Anal Chim Acta 500:185
35. Bro R (1996) Håndbog i multivariabel kalibrering. Jordbrugsforlaget, Copenhagen, p 53
36. Manne R (1987) Chemometr Intell Lab 2:187
37. Bro R, Smilde AK (2003) J Chemometr 17:16
38. Wroczynski RJ, Rabinsztajn M, Potyrailo RA (2004) Macromol Rapid Commun 25: 264
39. Hakkarainen M, Gallet G, Karlsson S (1999) Polym Degrad Stab 64:91
40. Lindström A, Hakkarainen M (2007) J Appl Polym Sci 104:2458
41. Capone S, Epifani M, Quaranta F, Siciliano P, Taurino A, Vasanelli L (2001) Sens Actuators B 78:174
42. Vallejo-Cardoba B, Nakai S (1994) J Agric Food Chem 42:989
43. Marsili RT (2000) J Agric Food Chem 48:3470
44. Di Natale C, Macagnano A, Martinelli E, Paolesse R, Proietti E, D'Amico A (2001) Sens Acutators B 78:26

45. Macagnano A, Carche M, Herrero A, Paolesse R, Martinelli E, Pennazza G, Carmano P, D'Amico A, Di Natale C (2005) Sens Acutators B 111:293
46. Sakakibara K, Kaitani K, Hamada C, Sato S, Matsuo M (1999) JSAE Review 20:237
47. Morvan M, Talou T, Gaset A, Beiau (2000) Sens Acutators B 69:384
48. Gerrigues S, Talou T, Nesa D, Gaset A (2001) Sens Acutators B 78:337
49. Morvan M, Talou T, Beziau J-F (2003) Sens Acutators B 95:212
50. Potyrailo RA, Wroczynski RJ, Morris WG, Bradtke GR (2004) Polym Degrad Stab 83:375
51. Lindström A, Albertsson A-C, Hakkarainen M (2004) Polym Degrad Stab 83:487
52. Hakkarainen M, Albertsson A-C (2005) Biomacromolecules 6:775
53. Gröning M, Hakkarainen M (2002) J Appl Polym Sci 86:3396
54. Gonzalez-Velasco JR, Gonzales-Marcos JA, Delgado JA, Gonzales-Ortiz De Elguea C, Gutierrez-Ortiz JI (1996) Chem Eng Sci 51:1113
55. Gröning M, Hakkarainen M (2001) J Chromatogr A 932:1
56. Albertsson A-C, Barenstedt C, Karlsson S, Lindberg T (1995) Polymer 36:3075
57. Albertsson A-C, Erlandsson B, Hakkarainen M, Karlsson S (1998) J Environ Polym Degrad 6:187
58. Hakkarainen M, Karlsson S, Albertsson A-C (1999) Polymer 41:2331
59. Hakkarainen M, Karlsson S, Albertsson A-C (2000) J Appl Polym Sci 76:228
60. Hakkarainen M, Albertsson A-C (2002) Macromol Chem Phys 203:1357
61. Marie A, Fournier F, Tabet JC (2000) Anal Chem 72:5106
62. Tuchbreiter A, Marquardt J, Zimmermann J, Walter P, Mulhaupt R (2001) J Comb Chem 3:598
63. Kappler B, Tuchbreiter A, Faller D, Liebetraut P, Horbelt W, Timmer J, Honerkamp J, Mulhaupt R (2003) Polymer 44:6179
64. Tuchbreiter A, Kappler B, Stockmann R, Mulhaupt R, Honerkamp J (2003) Macromol Mater Eng 288:29
65. Burman L, Albertsson A-C (2005) Polym Degrad Stab 89:50
66. Hakkarainen M, Albertsson A-C, Karlsson S (1997) J Appl Polym Sci 66:959
67. Hoff A, Jacobsson S (1981) J Appl Polym Sci 26:3409
68. Albertsson AC, Barenstedt C, Karlsson S (1994) Acta Polym 45:97
69. Arnaud R, Moisan JY, Lemaire J (1984) Macromolecules 17:332
70. Gugumus F (2002) Polym Degrad Stab 77:147
71. Hock H, Kropf H (1957) Angew Chem 69:313
72. Plesnicar B (1983) In: Patai S (ed) The Chemistry of Peroxides. Wiley, Chichester, p 521

Adv Polym Sci (2008) 211: 23–50
DOI 10.1007/12_2007_118

Published online: 17 August 2007

Solid Phase Microextraction for Analysis of Polymer Degradation Products and Additives

Minna Hakkarainen

Department of Fibre and Polymer Technology,
School of Chemical Science and Engineering,
Royal Institute of Technology (KTH), 10044 Stockholm, Sweden
minna@polymer.kth.se

Abstract Commercial plastics and rubbers always contain low molecular weight additives. Other low molecular weight compounds are formed due to degradation of polymers and additives during synthesis, processing, use and after disposal. Knowledge of these compounds is important for evaluating the environmental impact and interaction of polymeric materials, to investigate long-term properties and degradation mechanisms, to verify ingredients, to investigate manufacturing problems, to quality control polymeric materials, to identify odorants, to avoid workplace exposure and to insure safety of food packaging and medical products. Solid-phase microextraction (SPME) is a solvent-free extraction technique that has in recent years found an increasing number of applications in different fields. This paper presents a review of SPME technique in different polymer related applications including analysis of degradation products and polymer additives, monomer-rests, odor compounds, migrants from food packaging and medical products, extraction of polymer additives from environmental samples and biological fluids. Future possibilities are also discussed.

Keywords Solid phase microextraction · Polymer analysis · Degradation products · Additives

Abbreviations

BADGE	bisphenol A diglycylether
BBP	butylbenzyl phthalate
BFR	brominated flame retardant
BHT	butylated hydroxytoluene
BPA	bisphenol A
CAR	carboxen
CW	carbowax
DBP	di-*n*-butyl phthalate
DBT	dibutyltin
DEHP	di-2-ethylhexyl phthalate
DEP	diethyl phthalate
DMP	dimethyl phthalate
DNP	di-*n*-nonyl phthalate
DOP	di-*n*-octyl phthalate
DPA	dibutyl amine
DSC	differential scanning calorimetry
DVB	divinylbenzene
EPS	expanded polystyrene
FPD	flame photometric detection
FTIR	fourier transform infrared spectroscopy
GC	gas chromatography
GC-MS	gas chromatography-mass spectrometry
HPLC	high performance liquid chromatography
HS	headspace
LC	liquid chromatography
LDPE	low density polyethylene
LPME	liquid-phase microextraction
MAE	microwave assisted extraction
MBT	monobutyltin
MHE	multiple headspace extraction
MHS	multiple headspace
MS	mass spectrometry
NBBS	*N*-butylbenzenesulfonamide
PA	polyacrylate
PBB	polybrominated biphenyl

PBDE	polybrominated diphenyl ether
PCB	polychlorinated biphenyl
PCL-PC	polycaprolactone-carbonate
PDMS	polydimethylsiloxane
PET	poly(ethylene terephthalate)
PLLA	poly(l-lactide)
PMMA	poly(methyl methacrylate)
PS	polystyrene
PVC	poly(vinyl chloride)
RSD	relative standard deviation
SBR	styrene-butadiene rubber
SHS	static-headspace
SPME	solid phase microextraction
TBEP	tris(2-butoxyethyl)phosphate plasticizer
TBT	tributyltin
TMTD	tetramethylthiurame disulfide
TPR	templated resin
UV	ultraviolet
VOC	volatile organic compound

1 Introduction

1.1 Low Molecular Weight Compounds in Polymers

Commercial plastics and rubbers always contain a number of additives that are included to give particular physical and/or chemical properties. These additives include plasticizers, extender oils, carbon black, inorganic fillers, antioxidants, antiozonants, heat and light stabilizers, tackifying resins, processing aids, cross-linking agents, accelerators, retarders, adhesives, pigments, smoke and flame retardants, and others. New low molecular weight compounds are also formed due to the degradation of polymer matrix and additives during synthesis, processing and service-life as well as after disposal of the polymeric materials (Fig. 1). Knowledge of additives and degradation products is important to verify the ingredients, to reconstruct recipes of unknown materials, to investigate manufacturing problems, to identify odorants or irritants that evolve from polymeric materials and for evaluating the degradation mechanisms and long-term properties. Because of the environmental concern for chemicals, the total composition of polymeric materials has become an important issue. The manufacturing worker, the people involved with polymer processing and fabricating and finally the consumer should not be exposed to toxic substances. The composition of the material is especially important in applications such as food wrappings, utensils for eating and cooking, toys, and medical and biomedical products. Many polymer additives

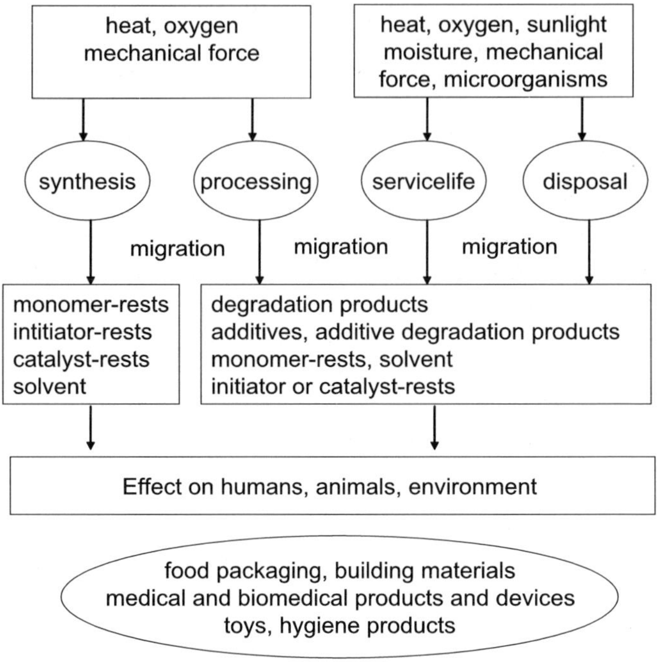

Fig. 1 Polymeric materials contain different low molecular weight compounds that are deliberately added or formed during the life-cycle of the material

are produced in large quantities and can be found in different environmental and biological samples.

1.2 Solid-phase Microextraction

Solid phase microextraction (SPME), introduced in the 1990s by Pawliszyn [1], is an inexpensive, rapid and solvent-free extraction technique that integrates sampling, extraction, concentration and sample introduction into a single step. It is based on a thin fused silica fiber coated with a polymeric stationary phase mounted in a modified syringe. During extraction, the analytes are adsorbed or absorbed by the fiber, depending on the type of stationary phase used. After equilibrium is reached or after a pre-defined time, the fiber is withdrawn and transferred into either an injection port of a gas chromatograph (GC) or a modified valve of a liquid chromatograph (LC). The analytes are thermally desorbed in the hot GC injector or eluted by mobile phase in the LC injector and transferred to the GC or LC column. Solid-phase microextraction is a versatile method that can be used for extraction of volatiles and semi-volatiles from gaseous, solid and liquid matrices. When extracting semi-volatiles from an aqueous matrix, the fiber is usually directly immersed into the sample. Agitation of the sample by a magnetic stirrer or by sonication greatly reduces the equilibrium time. Headspace (HS) sampling is used for more volatile compounds and has the advantage of a faster equilibrium time. HS sampling is also used for dirty or solid matrices.

SPME is a multiphase equilibrium technique and, therefore, the analytes are not completely extracted from the matrix. Nevertheless, the method is useful for quantitative work and excellent precision and linearity have been demonstrated. An extraction is complete when the concentration of analytes has reached distribution equilibrium between the sample and coating. This means that once the equilibrium is achieved, the amount extracted is independent of further increase in extraction time. If extraction is terminated before the equilibrium is reached, good precision and reproducibility is still obtained if incubation temperature, sample agitation, sample pH and ionic strength, sample and headspace volume, extraction and desorption time are kept constant. The theory of the thermodynamic, kinetic and mass transfer processes underlying direct immersion and HS-SPME has been extensively discussed by Pawliszyn [2]. The sensitivity and time required to reach adsorption equilibrium depends on the partition coefficients between the fiber and the analytes, and the thickness of the phase. Limits of detection and quantitation are often below 1 ppb.

There are several commercially available fiber materials with different polarities and fiber thicknesses and the selectivity of the extraction can be altered by changing the fiber material. Non-polar polydimethylsiloxane (PDMS) fiber is ideal for extraction of non-polar compounds. For extraction of volatiles, a thicker fiber should be chosen, while semi-volatiles are preferentially extracted by a thinner fiber. Polydimethylsiloxane/divinylbenzene (PDMS/DVB) fiber is designed for extraction of polar volatiles, carboxen/polydimethylsiloxane (CAR/PDMS) for extraction of gases and volatiles and DVB/CAR/PDMS for extraction of odors and flavors. Polar polyacrylate (PA) fiber is ideal for extraction of polar semi-volatiles such as phenols. Polar Carbowax/divinylbenzene (CW/DVB) is suitable for extraction of alcohols and other polar compounds. For HPLC, there are PDMS, PDMS/DVB, PA and carbowax/templated resin (CW/TPR) fibers. Both PDMS and PA phases consist of a high viscosity liquid that extracts analytes via absorption, i.e., the analytes dissolve and diffuse into the bulk of the coating. The remaining types are mixed coatings where the primary extracting phase is a porous solid that extracts via adsorption, i.e., the analytes stay on the surface of the fiber. In general, polar fibers are suitable for extraction of polar analytes and non-polar fibers for extraction of non-polar analytes. The fiber thickness affects the equilibrium time and sensitivity. Usually the thinnest film is employed in order to reduce the extraction time. Since its introduction, SPME has rapidly found numerous applications in environmental, food, pharmaceutical, clinical and forensic applications [3].The number of polymer related applications is also constantly increasing.

2 Polymer Degradation Products

The identification of polymer degradation products is essential for understanding the long-term properties and degradation mechanisms as well as the interaction of polymers with the environment, including possible toxicological or ecotoxicological effects. The degradation product pattern also gives information regarding the degree of oxidation in the polymer matrix [4, 5]. It can be used to differentiate between biotic or abiotic degradation [6, 7] or to detect early signs of degradation in the material. Product patterns have also been used to estimate the effectiveness of different antioxidant packages during processing [8].

2.1 Thermo- and Photo-oxidation Products

In several studies, SPME was applied for extraction of thermo- or photo-oxidation products from low density polyethylene (LDPE) and enhanced environmentally degradable polyethylene [9–12]. Hydrocarbons, ketones, furanones and carboxylic acids were the main products extracted by HS-SPME from enhanced environmentally degradable polyethylene thermo-oxidized at 60, 80 or 100 °C. In some cases, aldehydes and esters were also detected. Polar carbowax-divinylbenzene fiber effectively extracted ketones, furanones and carboxylic acids from thermo-and photo-oxidized polyethylene [9]. Non-polar PDMS was suitable for the extraction of ketones and furanones, however its extraction efficiency for polar carboxylic acids was low. Thermostatting temperature during the extraction was 80 °C and extraction time was 30 minutes. Lower extraction temperature was not enough to volatilize the low molecular weight compounds and higher temperature favored desorption of compounds from the fiber. As seen from Fig. 2, the number of products extracted from photo-oxidized LDPE by HS-SPME was several times larger compared to the number of products detected after traditional HS extraction. The SPME method allowed for the identification of a homologous series of carboxylic acids, ketones and furanones while only a few carboxylic acids and traces of ketones were detected after traditional headspace extraction [9, 13]. In both cases, the products were identified by gas chromatography-mass spectrometry (GC-MS). SPME was especially more effective in extracting less volatile products such as longer carboxylic acids and larger furanones. A series of ketones, linear and branched alkanes, alkenes, carboxylic acids, furanones, alcohols and esters were extracted from photo-oxidized polyethylene [9]. Several degradation products originating from the additives were also detected. The use of styrene-butadiene copolymer (SBR) in the pro-oxidant formulation for enhanced environmentally degradable polyethylene resulted in several aromatic degradation products, including benzoic

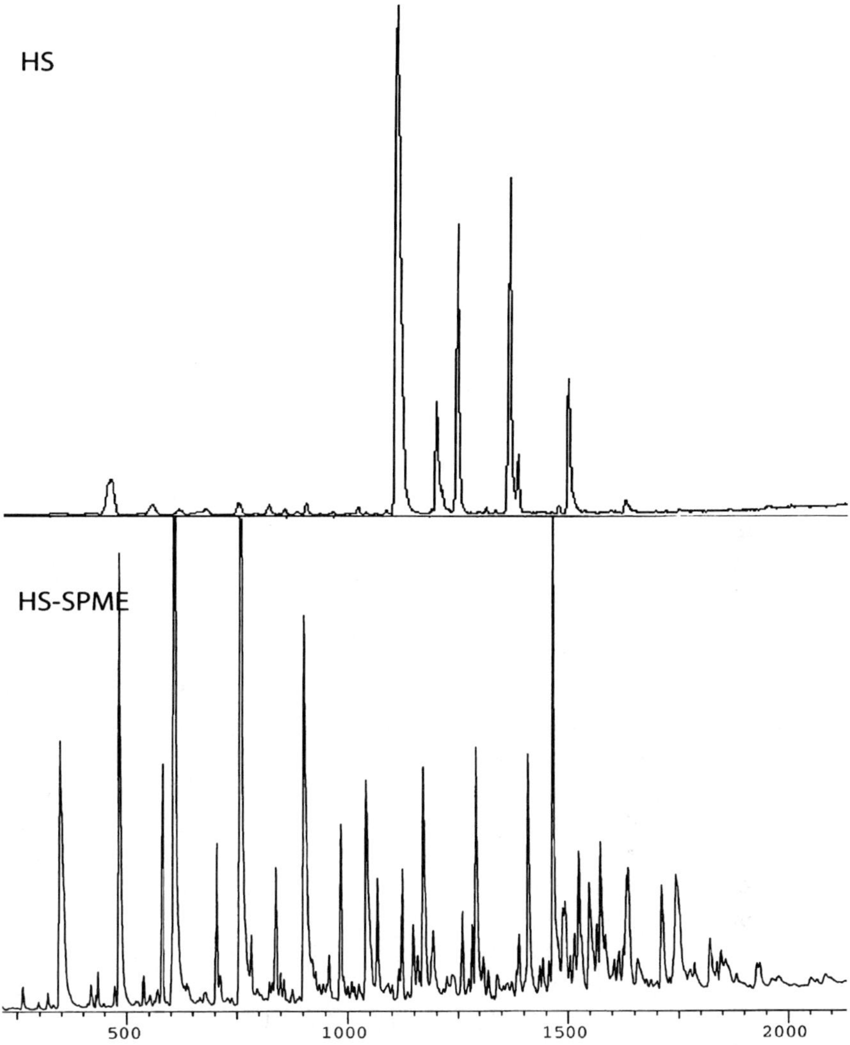

Fig. 2 GC-MS chromatograms showing the products extracted from photo-oxidized polyethylene by traditional HS extraction (*above*) and by HS-SPME with polar carbowax/divinylbenzene fiber (*below*). A few short-chain carboxylic acids were detected in the chromatogram after HS extraction, while carboxylic acids, ketones and furanones were detected after HS-SPME

acid, benzaldehyde, acetophenone, benzyl benzoate and two other benzene derivatives [12].

The HS-SPME method was also developed for the extraction of low molecular weight compounds from virgin or thermo-oxidized polyamide-66 [14, 15]. Figure 3 shows GC-MS chromatograms obtained after 25 and 1200 hours

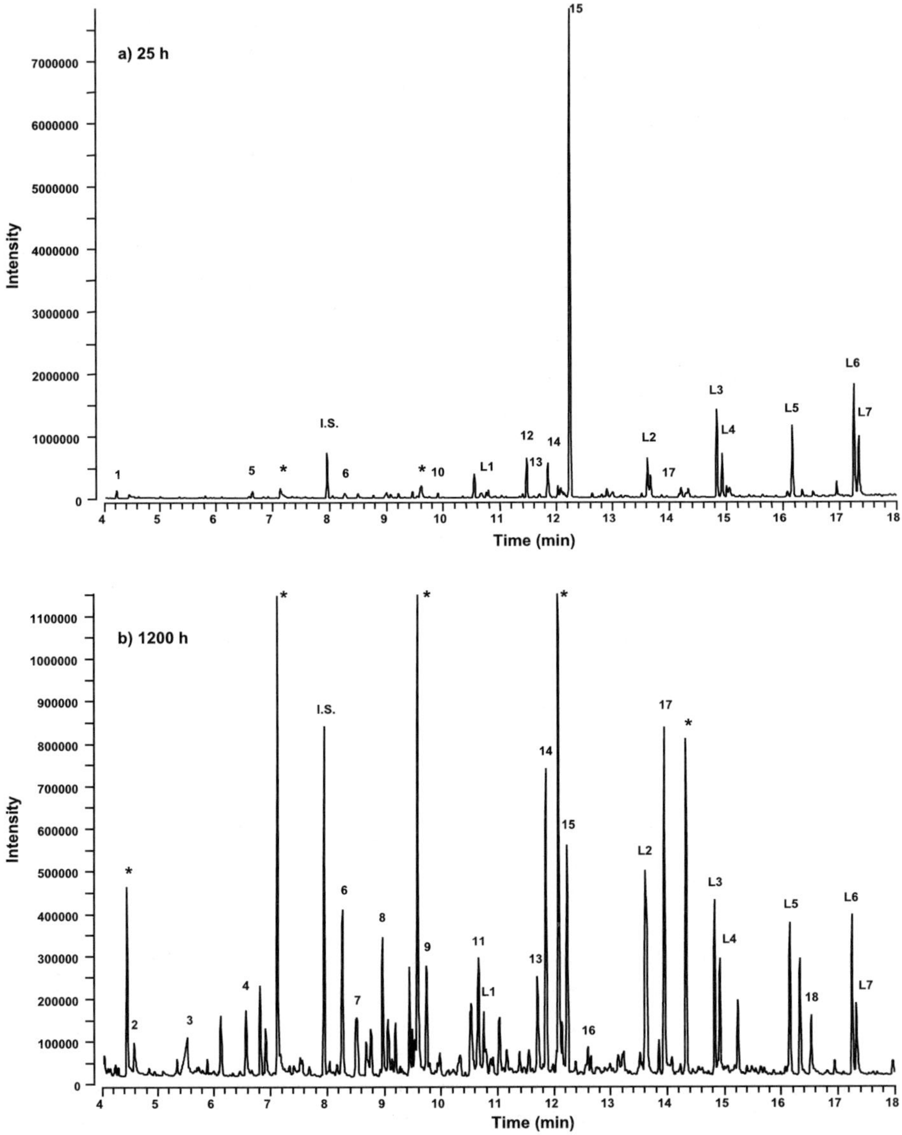

Fig. 3 GC-MS chromatograms after **a** 25 h and **b** 1200 h of aging of polyamide 6.6 at 100 °C. Peak 1–18 low molecular weight degradation products evolved from polyamide 6.6 including cyclic imides, pyridine derivatives, chain fragments and cyclopentanones. L1–L7 alkanes and alkenes from a lubricant added to the polymer. S.I. = internal standard. Reprinted from [14] with permission from Elsevier

at 100 °C. The low molecular weight compounds included cyclic imides, pyridines, different chain fragments and cyclopentanones. The effect of fiber material, extraction temperature and time, incubation time and sample size on the extraction were evaluated [16]. Traditional HS-extraction was more presice than HS-SPME, with the relative standard deviation (RSD) generally around 5%, while the RSD for manual HS-SPME was approximately 10%. However, the limit of detection was more than four times lower if compounds were extracted by HS-SPME instead of traditional HS-extraction. Quantitative headspace analysis of volatiles in solid matrices is complicated by interactions between the solid matrix and analytes. These matrix effects can be removed by using multiple headspace extraction, which can also be combined with SPME [17]. To remove the matrix effects between polyamide 6.6 and 2-cyclopentyl-cyclopentanone, a multiple headspace technique was developed for the quantitative determination of 2-cyclopentyl-cyclopentanone [18]. However, the strong hydrogen bonding between the analyte, 2-cyclopentyl-cyclopentanone, and the polar polyamide 6.6 caused slow migration rate and adsorption of analyte and led to erroneous quantification despite the use of multiple headspace extraction. This was solved by the addition of water, which disrupted the hydrogen bonding between the analyte and the matrix, and a valid quantification was finally achieved. The addition of water also increased the sensitivity.

The high sensitivity of SPME-GC-MS as a polymer analysis tool was also demonstrated as changes in the degradation product pattern were detected considerably earlier than tensile testing, differential scanning calorimetry (DSC) or fourier transform infrared spectroscopy (FTIR) could detect any changes in the polymer matrix [15]. At later stages, however, the recorded degradation product patterns correlated well with the simultaneous changes in mechanical properties and the largest increase in the number and amount of degradation products coincided with the time when rapid decrease in tensile strength was detected. A good correlation was observed between the amount of 2-cyclopentyl-cyclopentanone determined by microwave assisted extraction (MAE) and by MHS-SPME.

In an unpublished study, SPME and a 65 μm PDMS/DVB fiber were applied for extraction of polystyrene (PS) thermo-oxidation products. As seen in Fig. 4, styrene monomer was the main compound migrating from the virgin material after low temperature thermo-oxidation at 80 °C. After 20 weeks, new low molecular weight compounds including toluene, ethyl benzene, benzaldehyde and acetophenone were also identified. Volatiles migrating from poly(vinyl chloride)/polycaprolactone-carbonate (PVC/PCL-PC) during thermo-oxidation were extracted by PDMS/DVB fiber [19]. The identified compounds included degradation products of PCL-PC, i.e., 6-hydroxyhexanoic acid and caprolactone. CW/DVB fiber was used to extract thermo-oxidation products from poly(ethylene oxide-propylene oxide-ethylene oxide) [20]. The identified compounds mainly included different formates and acetates. Chambers et al. have developed SPME methods to study aging and defects in polymers while they are deployed in nuclear weapons [21]. Chlorofluoropolymers, polyurethanes and polysiloxanes were

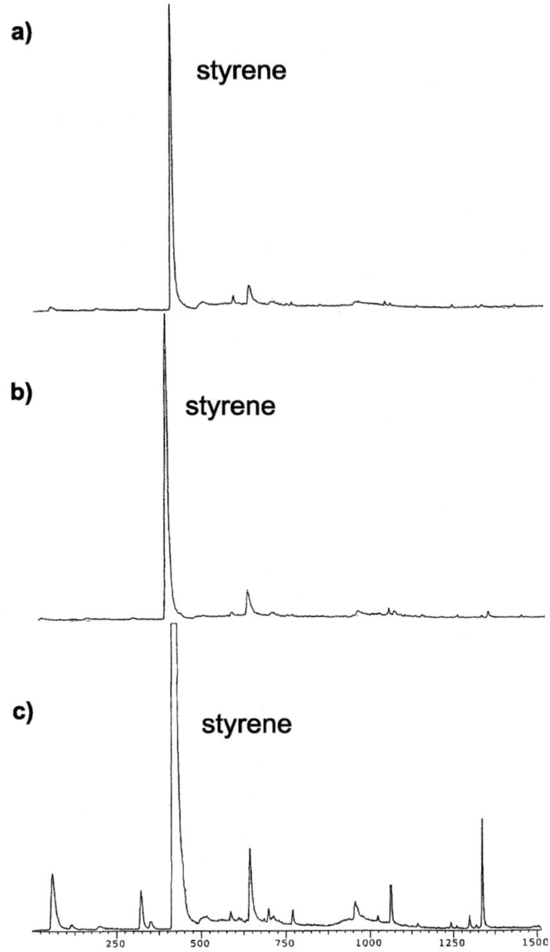

Fig. 4 Compounds migrating from low temperature thermo-oxidized polystyrene extracted by SPME after **a** 0 weeks, **b** 5 weeks and **c** 20 weeks at 80 °C. The most abundant peak in the chromatograms is styrene monomer

aged under environmentally relevant conditions and volatile and semi-volatile compounds emitted from the materials were collected by SPME.

2.1.1 Odor Compounds

Flexible multilayer packaging materials obtained by an extrusion coating process are widely used for food, cosmetics or medicines. The low molecular weight compounds formed during processing can impart undesirable odors and tastes to the content of the packaging. The odor- responsible compounds are usually different carbonyl compounds such as aldehydes, ketones and carboxylic acids. In addition to the undesirable organoleptic features, the low molecular weight compounds present in polymeric materials also affect properties such as flexibility, stability and strength. Therefore, analysis of these compounds in packaging materials is an important quality control tool to optimize the manufacturing conditions. Since the concentration of odor-responsible volatile organic compounds (VOC)s is usually very low, a sensitive method is needed to quality control the process. Common techniques are purge and trap and direct thermal desorption techniques followed by GC-MS identification. Tena et al. have developed SPME methods to identify and quantify odor-causing volatile compounds formed due to the thermo-oxidative degradation of flexible polymer packaging during extrusion coating [22, 23]. The packaging material consisted of a layer of cellulose, clay-coated paper or satin cellulose, a layer of polyethylene, a layer of aluminium and another layer of polyethylene. Twenty-five compounds were extracted by CAR/PDMS fiber and identified including aldehydes, ketones and carboxylic acids all known as odor causing compounds. The developed SPME method surpassed the sensitivity of static headspace extraction. The highest amount of carbonyl compounds was found in the packaging with unacceptable odor.

In order to quantitate the low molecular weight compounds, the authors also compared external calibration, standard addition and multiple headspace extraction techniques [23]. The choice of the calibration method had a large impact on the obtained results. The results clearly demonstrated the difficulty in the quantitative determination of volatiles in solid matrices and the importance of the right calibration method. Especially, the use of external standard calibration without taking the matrix-volatile interactions and changes in the headspace volume into account resulted in extremely high errors in the results, i.e., the calculated amounts were up to 10–50 times larger than if the effect of the polymer matrix was simulated by the addition of hexadecane or when standard addition or multiple headspace extraction (MHE) were applied. CAR/PDMS fiber was better for the extraction of the most volatile compounds, while DVB/CAR/PDMS was better for compounds with somewhat higher molecular weights. A multiple headspace solid phase microextraction method was developed and evaluated for the quantification of volatiles from four multilayer packaging samples [24]. In this method, hexadecane was used as a solvent for the preparation of calibration solutions. Later, the MHS-SPME method was further developed and hexadecane was replaced by water [25]. This reduced the extraction time and the number of extractions needed. It also improved reproducibility and linearity coefficients. The total peak areas obtained by MHS-SPME in hexadecane solutions were statistically equal to those obtained in aqueous solutions. This supports the conclusion that matrix effects were removed by the developed MHS-SPME method. Recently, HS-SPME was also applied to identify cosmetic ingredients that cause delamination of multilayer packaging materials [26]. HS-SPME has also been developed and applied as a rapid quality control tool for the multilayer packaging manufacturing process [27]. The quantification of volatiles in packaging samples manufactured under different conditions was carried out using a sample with a known amount of VOCs as a reference. The amount of VOCs in the reference sample was determined by the MHS-SPME method developed earlier by the same authors [24]. The new method reduced the analysis time from 5 h/sample (multiple HS-SPME) to 1.25 h/sample, which is a significant improvement.

Odor compounds may also be released from the plastic materials used in cars. The variety of plastics and possible chemical compounds is broad, which makes the identification of odor causing compounds an extremely complicated task. An effective and rapid screening of VOCs and semi-VOCs from materials used in automobiles was developed by utilizing the SPME technique [28]. The low molecular weight compounds extracted from five different automobile materials included different benzene derivatives, aldehydes, esters, biphenyls, phthalates, butylated hydroxytoluene, phenols, alcohols, styrene, triethylenediamine, carboxylic acids and ketones. A considerable number of VOCs and semi-VOCs were detected, indicating that more attention should be paid to the selection of materials and additives for automotive parts.

2.2 Thermal Degradation Products

SPME was applied for the extraction of thermal degradation products evolved from polypropylene that was thermally degraded at 470 °C under nitrogen atmosphere [29]. More than 30 products including different alkenes, alkanes and alkadienes were identified after extraction with PDMS or carboxen/PDMS fiber. The SPME technique offered rapid analysis and great efficiency in pre-concentration of the products. In another study, the thermal stability of foamed polysiloxane rubber was evaluated by HS-SPME-GC-MS [30]. Also in this study a non-polar PDMS SPME fiber was used. Extraction time and temperature were 30 minutes and 50 °C. Different dimethylsiloxane cyclic species were extracted. How the level of these species change with the age of the material and the likely degradation processes that contribute to their production were evaluated. In most cases, the concentrations were only a few parts per million. 2-ethylhexanoic acid, a by-product from the hydrolysis of the tin-catalyst, was also detected. Another study aimed to use HS-SPME-GC-MS as a diagnostic tool for the early detection of polymer degradation [31]. HS-SPME-GC-MS with carboxen/PMDS fiber showed different degradation patterns for thermally aged polysulfide depending on if it was aged in the presence of Viton A (vinylidene-hexafluoropropene copolymer) or not. Polysulfide control samples exhibited almost no mass loss and the dominating degradation product was 1,3,6,7-dioxadithionane. However, new cyclic degradation products and considerable mass loss were observed for the polysulfide exposed to viton A already after 24 hours at 70 °C, indicating an acid-catalyzed degradation mechanism in the presence of Viton A. The extraction time and temperature were 20 minutes at 50 °C. Methyl methacrylate and several polymer additives were extracted by SPME after thermal aging of poly(methyl methacrylate) [32].

2.3 Degradation Products of Degradable Polymers

Environmental concern has led to an escalated interest in using biodegradable or environmentally degradable polymers as alternatives to traditional commodity plastics in, e.g., packaging and other single-use applications which can then be disposed by biological treatment such as composting. Plastic films used for agricultural applications are directly integrated into the soil, while compostable materials must pass through a biological treatment (i.e., composting). In both cases, fertile soil used for agriculture is the final environment. It is of primary importance to assure that the biodegradable materials do not accumulate in the soil and do not release toxic compounds that could hinder plants, animals and human beings by entering the food chain. SPME has in several studies been applied to the extraction of low molecular weight compounds from different soil samples [33] and to the extraction of intermediates produced during biodegradation of environmental contaminants [34]. As degradation intermediates are often formed in trace amounts and may have short lifetimes, the sample preparation technique needs to be rapid and sensitive in order to avoid loss of valuable information on the transformation pathways. Significant improvements in both speed and sensitivity are obtained by using the SPME techniques. SPME could, thus, be a valuable technique for the isolation of degradation intermediates formed during, e.g., composting of degradable plastics. Thus far, SPME has been applied for the extraction of low molecular weight compounds from poly(l-lactide) (PLLA) films after aging in soil for periods of up to two years [35]. Varying amounts of lactide, lactic acid and lactoyl lactic acid were extracted from the films by CW-DVB fiber depending on the soil burial time. The SPME technique was also applied for the extraction of hydrolysis products, such as 6-hydroxyhexanoic acid, after aging PVC/PCL-PC in aqueous environment [36]. Table 1 presents SPME methods developed for the extraction of degradation products from different polymeric materials.

3 Monomer and Solvent Rests in Polymeric Materials

Several SPME methods have been developed for the determination of residual solvents and monomers in polymeric materials. The amount of styrene

Table 1 Application of solid-phase microextraction for extraction of degradation products from polymers

Refs.	Polymer and aging	Extracted compounds	Fiber
Bortoluzzi, Pinheiro, Carasek, Soldi [29]	Polypropylene	Alkenes, alkanes, alkadienes	PDMS CAR-PDMS
Ezquerro, Pons, Tena [22–25, 27]	Multi-layer packaging thermo-oxidation during extrusion coating process	Ketones, aldehydes, carboxylic acids, hydrocarbons	CAR-PDMS
Gallet, Erlandsson, Albertsson, Karlsson [20]	Poly(ethylene oxide-propylene oxide-ethylene oxide) thermo-oxidation	Formates, acetates, carboxylic acids, aldehydes	CW-DVB
Gallet, Lempiäinen, Karlsson [35]	Poly(L-lactide) soil burial	Lactide, lactic acid lactoyl lactate	CW-DVB
Gröning, Hakkarainen [14–16, 18]	Polyamide 66 thermo-oxidation	Cyclopentanones, pyridines, cyclic imides, amides, carboxylic acids, caprolactam	PDMS-DVB
Hakkarainen, Albertsson, Karlsson [9]	Polyethylene Thermo-oxidation	Ketones, carboxylic acids, furanones, ketoacids	CW-DVB PDMS
Hakkarainen [19, 36]	Poly(vinyl chloride)/ polycaprolactone-carbonate blend Thermo-oxidation, hydrolysis	6-Hydroxyhexanoic acid, caprolactone	PDMS-DVB
Hall, Patel [30]	Polysiloxane rubber thermal degradation	Cyclic oligomers	PDMS
Khabbaz, Albertsson, Karlsson [10, 11] Khabbaz, Albertsson [12]	Polyethylene thermo-oxidation photo-oxidation	Hydrocarbons, ketones, esters, carboxylic acids, ketoacids, furanones, alcohols, aldehydes	CW-DVB PDMS
Rogalewicz, Voelkel [32]	Poly(methyl methacrylate) thermal degradation	Methyl methacrylate, additives	PDMS
Vance, Alviso, Harvey [31]	Polysulfide thermal degradation	1,3,6,7-dioxadithionane, other cyclic products	CAR-PDMS

monomer in polystyrene that was heated for different times and drawn into different shapes during the manufacturing process has been determined by HS-GC-MS [37]. In addition to styrene acrylonitrile, *t*-butylbenzene,

α-methylstyrene and butylated hydroxytoluene were detected. The manufacturer wanted to identify the volatiles in the polymer and to monitor the differences in the composition of the volatiles resulting from the variations in the process. The results from static-headpace (SHS) analysis were compared with SPME-GC-MS. With heated SHS, the recovery was biased towards the more volatile compounds, while the SPME gave higher recoveries for relatively nonvolatile compounds. SPME fiber was PDMS 100 μm. Extraction was performed for 45 minutes at room temperature. Kusch and Knupp analyzed residual styrene and other volatile organic compounds in expanded polystyrene (EPS) by headspace-SPME-GC-MS [38, 39]. EPS is used for food packaging and for protection of products against damage during transport and storage. It is also used in the building industry for isolating exterior walls. EPS emitted residual monomer and other volatile compounds at ambient temperatures. In addition to styrene monomer, pentane (blowing agent), benzene, toluene, ethylbenzene, isomers of xylene, *n*-propylbenzene, 1,2,4-trimethylbenzene, *o*-methylstyrene, benzaldehyde, benzyl alcohol and acetophenone were identified. Internal calibration was applied to quantify the amount of styrene monomer in the samples. They reported good reproducibility with RSD values between 3.2–3.6% using 75 μm carboxen-PDMS fiber. The HS-SPME-GC method proved to be a rapid and simple procedure to process control the EPS production.

HS-SPME-GC-MS showed that residual methyl methacrylate is released during thermal annealing of PMMA [40]. SPME fiber was carboxen/PDMS. The extraction time and temperature was 2 hours at 70 °C. A multiple headspace extraction method using a carboxen/PDMS fiber was developed for the quantitative determination of vinyl chloride monomer in PVC [41]. To reduce the equilibrium time, the PVC sample was finely ground before the extraction. Quantitative SPME methods have also been developed to determine vinyl chloride in liquid and solid samples [42] and to determine terephthalic acid and vinyl acetate monomers from aqueous solutions [43].

4 Polymer Additives

Long-term properties of polymers are severely affected and the service-life is reduced due to the migration of additives. There is also the possibility that some of the additives accumulate in the environment and affect our health and the environment. SPME has been applied for the extraction of several common polymer additives. Since additives often have rather low volatility, a significant advantage with SPME and HS-SPME compared to HS-GC-MS is the ability to extract even semi-volatile compounds [9, 13].

4.1 Plasticizers

4.1.1 Phthalate Esters

Phthalate esters are produced all over the world in large quantities for different industrial uses. One important application is the use as a polymer plasticizer. The release of phthalates into the environment may occur during production and distribution and due to migration from polymeric materials. Due to the widespread use, they have become common organic pollutants. In recent years, considerable attention has been paid to human exposure to phthalates because they are suspected to cause various health effects and possess carcinogenic and estrogenic properties [44–46]. Analysis of phthalate esters is often problematic due to high blanks caused by the presence of phthalates in many laboratory products, including chemicals and glassware. The use of SPME reduces the risk for secondary contamination.

Figure 5 shows a GC chromatogram with a large di-2-ethylhexyl phthalate (DEHP) peak obtained after HS-SPME of commercial PVC-tubing. Thermostatting time and temperature were 30 minutes at 80 °C. An ultrasonic sol-

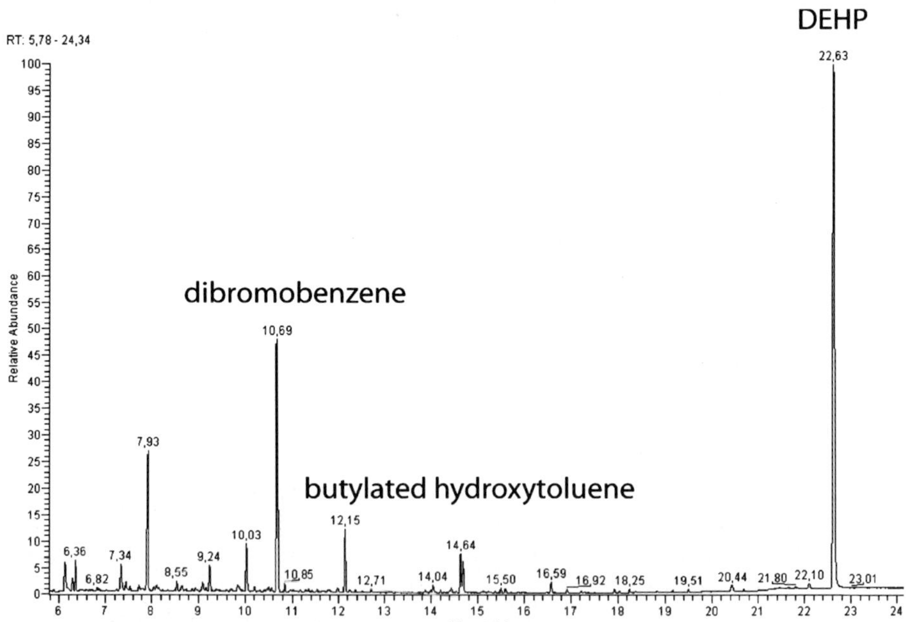

Fig. 5 GC-MS chromatogram showing the low molecular weight compounds extracted from commercial PVC tubing by HS-SPME. The most abundant peak in the chromatogram is DEHP

vent extraction method combined with SPME with calix [4]arene/ hydroxyl-terminated silicone oil coated fiber was developed and validated for the extraction of phthalate acid esters from plastic products such as blood bags, transfusion tubings, food packaging bags and mineral water bottles [47]. The correlation coefficients were better than 0.996 and relative standard deviations were $\leq 10\%$. Recovery of DEHP was between 95.5–101.4%. Several solid phase microextraction methods were developed for extraction of phthalates from different water samples [48, 49]. The SPME method was developed for the determination of diethyl phthalate (DEP) in water by using CW/TPR and PDMS/DVB fibers in combination with high performance liquid chromatography (HPLC) [50]. The limit of detection was $1\ \mathrm{ng\,mL^{-1}}$. The developed SPME method was applied for extraction of phthalates from various water samples including mineral water stored in plastic bottles. Six different phthalates and an adipate ester were determined in water samples by SPME-GC-MS method using a polyacrylate fiber [51]. The analyzed samples included tap and mineral water as well as water from a river, fishing and industrial ports. The limits of detection were $6–170\ \mathrm{ng\,L^{-1}}$.The miniaturized fiber-in-tube solid-phase microextraction method coupled with an LC system allowed the quantification of low levels of phthalates in waste water samples [52]. Hollow-fiber liquid phase microextraction (LPME) and SPME were compared for the extraction of different phthalates from water samples including bottled mineral waters [53]. Both techniques could be recommended for trace analysis of phthalates in drinking water and both techniques also eliminated the secondary contamination during analysis which is a frequent problem in phthalate analysis. In most cases, the detection limits varied from 0.005 to $0.1\ \mathrm{\mu g\,L^{-1}}$. In another study, the presence of 5 phthalate esters, bisphenol A and bisphenol A diglycyleter (BADGE) in distribution and bottled water was investigated [54]. Dimethyl phthalate (DMP), DEP, nonylphenol, butylbenzyl phthalate (BBP) and DEHP were extracted from water samples that had been stored in contact with plastic or painted concrete reservoirs and pipes. Multivariate optimization of the solid-phase extraction method for analysis of phthalate esters in environmental water was performed [55]. The obtained detection limits were in most cases in the low $\mathrm{pg\,L^{-1}}$ level and the developed method was successfully applied to several real environmental water samples including mineral, river, industrial port and sewage water samples. Phthalate esters were found in all the studied samples.

Luks-Betlej et al. evaluated six different fiber materials for determination of seven different phthalates in water samples [56]. The tested phthalates included DMP, DEP, di-*n*-butyl phthalate (DBP), BBP, DEHP, di-*n*-octyl phthalate (DOP) and di-*n*-nonyl phthalate (DNP). All fibers investigated were favorable for the extraction of BBP. Low molecular weight phthalates (DMP, DEP, DBP) were best extracted by DVB/CAR/PDMS, DBP by CW/DVB and DEHP by PDMS fiber. CW/DVB fiber gave the best repeatability. Application of CW/DVB fibers to different water samples allowed the detection of phthalates in low $\mu g\,L^{-1}$ concentrations in drinking water. DBP and DEHP were extracted in water soluble form in sewage sludge and river sediments by SPME coupled with HPLC-electrospray ionization-MS method with CW-DVB fiber [57]. Several phthalates were extracted by headspace and immersion SPME from the leachate from municipal solid waste landfills [58]. SPME was also applied to in vitro skin permeation experiments in order to obtain information on transport of diethyl phthalate through the skin [59]. SPME offered several advantages, being a fast and easily automated technique with adequate sensitivity and good precision. The results indicated that skin may be a significant route for the uptake of phthalate esters.

4.1.2 Organophosphate Triesters

Semi-volatile organophosphate triesters are commonly used as flame retardants and plasticizers in polymeric materials. HS-SPME-GC-MS was applied to follow the migration of tris(2-butoxyethyl) phosphate (TBEP) plasticizer and its degradation product, 2-butoxyethanol, as well as migration of other low molecular weight compounds from nitrile rubber during long-term aging [13]. During aging at 60 °C, the amount of 2-butoxyethanol increased slowly, but only a couple of new low molecular weight products were formed during 224 days of aging. A large increase in the number and amount of products migrating from the material was, however, seen when the temperature was raised from 60 to 80 °C. Most of these volatiles were degradation products of two additives in the material, i.e., tris(2-butoxyethyl)phosphate and tetramethylthiurame disulfide (TMTD). The identified products include a large amount of 2-butoxyethanol and smaller amounts of, e.g., *N*,*N*-dimethylformamide, dimethylurea and tetramethylurea. The TBEP plasticizer and some of the TMTD degradation products that were extracted and identified in the present study by HS-SPME-GC-MS were not detected in an earlier HS-GC-MS study of the same nitrile rubber even though a higher extraction temperature (100 °C vs. 80 °C) was used during the HS-GC-MS analysis [60]. This shows the ability of HS-SPME to extract even the rather non-volatile compounds present in solid materials. The solid phase microextraction method was also developed for the determination of organophosphate triesters from air samples under non-equilibrium conditions [61]. Detection limits of lower than 2 $ng\,m^{-3}$ for individual organophosphate esters were obtained. PDMS fiber had the highest uptake.

4.1.3 *N*-butylbenzenesulfonamide Plasticizer

A rapid SPME method was developed for the determination of *N*-butylbenzenesulfonamide (NBBS) plasticizer used, e.g., in polyamide and polyamide copolymers [62, 63]. Polyacrylate fiber was applied for extraction of NBBS from waste water. Sampling from water was performed at ambient temperature for 30 minutes under magnetic stirring. The limit of detection was 0.1 $\mu g\,L^{-1}$. NBBS was also detected in several raw and treated waste water samples from municipal waste water plants in the range from 0.1–3.5 $\mu g\,L^{-1}$.

4.2 Flame-retardants

Brominated compounds have been used as flame retardants in electronic equipment, textiles and building materials since the early 1970s after the prohibition of polychlorinated biphenyls (PCBs). However, there are also toxicological concerns due to the increasing amount of brominated flame retardants (BFRs) found in humans and in environmental samples [64]. SPME in combination with GC-MS has been used to screen for different brominated flame retardants in polyamide materials collected for recycling [13]. HS-SPME-GC-MS revealed the presence of several brominated compounds in recycled polyamide 6.6, all possible degradation products of the commonly used flame-retardant tetrabromobisphenol A. A quantitative analysis using 1,4-dibromobenzene and bromobenzene as model compounds for the identified tetrabromobisphenol A degradation products showed that the amounts of brominated compounds migrating from the recycled polyamide 6.6 ranged from 3 to 10 $\mu g\,kg^{-1}$ polyamide per analyte. The HS-SPME method was also able to extract fragments and degradation products from tetrabromobisphenol A and other brominated flame-retardants. Furthermore, the high extraction capacity of the PDMS/DVB stationary phase towards aromatic compounds was demonstrated, as the HS-SPME-GC-MS method allowed the extraction and identification of brominated benzenes from a complex matrix containing only trace amounts of analytes. In addition to the brominated compounds, degradation products from a hindered phenol antioxidant were extracted from the polyamide collected for recycling.

Solid-phase microextraction with PDMS fiber was applied for extraction of polybrominated diphenyl ethers (PBDEs) and polybrominated biphenyls (PBBs) from water samples [65]. By using the SPME technique, several cleanup steps could be avoided and the sample preparation time was greatly reduced. Detection limits varied from 7.5 to 190 $pg\,L^{-1}$. An in-situ acetylation-HS-SPME-GC-MS method was developed for the simultaneous determination of brominated phenols and other phenolic flame retardants including 3,5,3′,5′-tetrabromobisphenol A, 2,4-dibromophenol, 2,4,6-tribromophenol and pentabromophenol [66]. A multi-factor experimental design was created to study the main parameters affecting the extraction efficiency. Depending on the compound, the best results were obtained with CAR/PDMS or PDMS fibers. The method was also applied to several real samples including tap water and effluent and influent waste water from urban treatment plant in which several phenolic compounds could be detected.

Photo-degradation is suspected to be the main degradation pathway of BFRs in the environment. Therefore, it is of interest to know the photochemical behavior of BFRs in order to predict their fate in water, soil and plants, and to identify the photoproducts which could have biological and toxicological properties. The photo-degradation of polybrominated diphenyl ethers was investigated by utilizing an on-fiber SPME technique also called photo-SPME [67, 68]. Several polybrominated diphenyl ethers were extracted from aqueous solution by SPME. The BFRs were subsequently exposed to ultraviolet (UV) irradiation using the SPME fiber as photolytic support. The photoproducts formed in the fiber were identified and coincided with the photoproducts detected in aqueous photo-degradation experiments. Photo-SPME was also applied to study the degradation of tetra- to hexabrominated diphenyl ethers during exposure to natural and simulated sunlight [69]. The photochemical behavior in the fiber mimicked the process taking place in water under sun radiation. Finally ultrasound-assisted extraction was combined with SPME for the analysis of PBBs and PBDEs in sewage sludge samples [70]. The proposed method exhibited good performance in terms of linearity and precision, with recoveries exceeding 92% and the limits of detection were in the sub ng g^{-1} level. PBDEs were also detected and quantified from real contaminated sewage sludge and sediment samples. The photo-SPME method resulted in large solvent- and time-savings compared to the classic multi-step solvent extraction methods. Brominated flame retardants were also extracted from solid environmental samples such as soil and sediment [71]. Solid samples were moisturized with water and headspace extraction was performed with PDMS fiber for 60 minutes at 100 °C. The detection limits were in the sub-ng g^{-1} range.

The main advantage of the photo-SPME technique is that the generation of photoproducts takes place "in-situ" on the SPME fiber and no additional steps are needed to extract the generated products. The selectivity of the extraction technique often plays an important role in the obtained information. When the photolysis takes place directly on the fiber, all the generated photoproducts as well as original model compounds remain trapped on the fiber and the whole system can be analyzed simultaneously. The obtained relative ratios are thus not influenced by, e.g., different extraction efficiencies for different compounds. This provides unique qualitative and quantitative information regarding the degradation mechanisms and has great potential in evaluating the photo-transformation of organic compounds including different polymer additives.

4.3
Extraction of Polymer Additives from Biological Fluids

The analysis of body fluids can be used to monitor the exposure to different environmental or workplace pollutants. HS-SPME is ideally suited for the analysis of biological samples. It has high sensitivity and cleaner extracts are achieved due to reduced interference from high molecular weight species. Phthalic mono-and diesters were determined in urine samples by utilizing SPME and diazomethane on-fiber derivatization [72]. PDMS/DVB fiber gave the best results and allowed the detection of trace level concentrations (0.3–8.6 ng mL^{-1}). SPME was also applied to extract 4-heptanone, which is a β-oxidation product from 2-ethyl hexanoic acid from DEHP, from urine samples [73]. The SPME method based on a polyacrylate fiber was developed to detect and quantify chlorinated bisphenol A (BPA) in human plasma in order to assess human exposure [74]. Bisphenol A is used in the manufacturing of epoxy and phenolic resins, polycarbonates and corrosion resistant unsaturated polyester-styrene resins. BPA that migrates from plastics to tap water can react with chlorine to produce chlorinated BPA, which is more cytotoxic than BPA [75]. Isocyanates are widely used in the production of polyurethanes and constitute a significant workplace hazard [76]. To measure the workplace exposure, a personal isocyanate sampler was developed by loading dibutylamine (DBA) onto PDMS/DVB SPME fiber. The DBA-isocyanate derivatives were then desorbed by sonication and analyzed by LC-MS.

5
Migration from Polymeric Food Contact Material

Migration from plastic packaging into food can affect the safety and acceptability of the food. SPME has been applied for the extraction of antioxidants and their possible degradation products from simulated food and pharmaceutical solutions [77]. In another study, deliberately contaminated poly(ethylene terephthalate) (PET) samples were coated with a functional barrier to study the possibility of preventing the migration of contaminants from the recycled materials to food by using a layer of virgin material as a migration barrier [78]. Migration tests were then performed with different food stimulants and the migrants were extracted from the food simulants by SPME. DVB/CAR/PDMS fiber was used to extract the low molecular weight compounds migrating from plastic-based nets to water under refluxing conditions [79]. In addition to studying the migration of packaging components to food, SPME has also been applied to study the beneficial or adverse effects of packaging material on storage stability and aroma compounds in foods [80].

5.1 Acetaldehyde and Terephthalic Acid in PET Bottles

In a couple of studies, SPME methods were developed to follow the migration of low molecular weight compounds from bottles for mineral water. Acetaldehyde is a common degradation product of poly(ethylene terephthalate) formed during the melt condensation reaction and melt processing of PET. A rapid and sensitive SPME method was developed to extract acetaldehyde by inserting the carbowax/divinylbenzene fiber into the inner air-space of PET bottles [81]. The detection limit was 0.5 mL L^{-1} and relative standard deviation was lower than 7%. The acetaldehyde content of 50 PET bottles was analysed. In another study, the SPME method was developed to determine the terephthalic acid monomer from aqueous solutions [43].

5.2 Butylated Hydroxytoluene in Bottled Drinking Water

Butylated hydroxytoluene (BHT) is an antioxidant commonly utilized as an additive in food and in rubber and plastic products. Its chemical migration from plastic packaging into the package's content has been shown [82]. A method for the determination of BHT in bottled water by means of SPME-GC-MS was developed [83]. The method was also applied to evaluate the presence of BHT in mineral and mineralized bottled drinking water and it appeared in seven out of fifteen samples studied. The maximum observed concentration was 38 $\mu g\,L^{-1}$. All studied plastic bottles were made of PET. The PDMS fiber was exposed to 15 mL aliquots for 30 minutes at ambient temperature and under magnetic stirring. Relative standard deviation was 4.3% and the limit of detection was 4.2 $\mu g\,L^{-1}$.

5.3 Bisphenol-type Contaminants

Bisphenol A is used as a raw material to make polycarbonate and epoxy adhesives and can coatings. It is also used in flame-retardants, in unsaturated polyesters and in polyacrylate resins. Many foodstuff containers are made of these resins, including containers for oven and microwave cooking. Recent studies have shown that bisphenol type compounds have both mutagenic and cytotoxic properties [84]. Nerin et al. developed a fast screening method based on SPME and HPLC with fluorescence detection suitable for the analysis of several bisphenol derivatives and their degradation products in aqueous canned foods such as tuna, olives and corn [85]. The best results were obtained with carbowax and PDMS/DVB fibers. The detection limits were between 0.7 and 2.4 ng mL^{-1}, while RSDs were between 14 and 32%. After the extraction parameters were optimized, the method was applied to the analysis of several commercially canned samples. The conclusion was that SPME is a powerful and fast technique for the direct screening of non-volatile migrants in aqueous foodstuffs. A method for simultaneous analysis of bisphenol A and bisphenol A diglycidyl ether in aqueous food simulants was developed by utilizing direct immersion SPME and gas chromatography-mass spectrometry [86]. The detection limits by using polyacrylate fiber ranged from 0.1 to 2.0 ng g^{-1} for BPA and from 13 to 15 ng g^{-1} for BADGE. An automated on-line in-tube solid phase microextraction method was developed for the analysis of bisphenol A in foods in contact with plastics [87]. The developed method was easily automated and reduced solvent use and could be used both for analysis of food samples and migration tests from plastics. Solid-phase microextraction in combination with GC-MS showed that 0.7–78.5 μg L^{-1} of bisphenol A leached from plastic containers to water [88]. The detection limit could be lowered from 2 ng L^{-1} to 0.4 ng L^{-1} by derivatization with bis(trimethylsilyl) trifluoroacetamide.

5.4 Phthalate Esters

The presence of phthalates has decreased in food-packaging materials. However, they are still used in the food packaging industry as adhesives, offset printing inks and lacquers. An automated on-line in-tube solid phase microextraction method was developed for the analysis of phthalate esters and alkylphenols such as nonylphenol in foods in contact with plastics [87]. Several phthalates were also extracted by SPME from mineral water stored in plastic bottles [47, 50, 51, 53, 54]. In the modern dairy farm, milking is done by a machine system employing flexible tubing of various polymers. Milk samples were collected from the same cows by hand milking and by machine milking with plasticized PVC tubing [89]. The HS-SPME showed 10–20 times higher DEHP contents in the milk collected by machine milking. Analysis of the PVC tubing material indicated that the tubing contained 28 wt. % of DEHP.

5.5 Butyltin Compounds in Beverage and Food Packaging

Butyltin compounds, monobutyltin (MBT) and dibutyltin (DBT) and tributyltin (TBT) are widely used for the production of biocides and polymer stabilizers. On a yearly basis, 23, 000 tons of butyltin compounds are used as additives in PVC products. This is about 40% of the world usage of organotin compounds [90]. The negative impact of butyltins on the aquatic environment is well documented and in vitro studies have also shown that they disrupt the immune response in humans [91]. Studies have shown that organotins leach out from PVC and related materials, resulting in the contamination of foodstuff and beverages [92]. HS-SPME with PDMS fiber and gas chromatography with flame photometric detection (GC-FPD) was applied to determine butyltin and octyltin stabilizers in PVC products [93]. PVC was first dissolved in tetrahydrofuran. The stabilizers were hydrolysed to chloride forms by treatment with HCl and then derivatized with sodium tetraethylborate. Recovery was $> 90\%$ for butyltin and $> 80\%$ for octyltin compounds. SPME-GC-MS methods were developed to make a survey of the presence of butyltin compounds in portuguese [94] and chinese wines [95]. In the first study, derivatization with sodium tetraethylborate was applied, while potassium tetrahydroborate was used in the second study. A PDMS fiber was used for the extraction in both studies. Measurable amounts of DBT were found in 14% of the studied portugueses wines and MBT was found in one wine. RSDs were 8–9% and limit of detection varied between 0.01 and 0.2 $\mu g\,L^{-1}$ of Sn depending on the butyltin compound and type of wine. The results suggested that high-density polyethylene containers used in the transfer of wine in an early stage of the vinification were the main source of contamination. Sampling wines in PVC-lined storage tanks also caused high butyltin levels.

6
Migration from Pharmaceutical Packaging and Medical Materials

Registration of pharmaceutical and biomedical products requires extensive testing to ensure the safety of the finished product. This includes the physical and chemical stability of the drug formulation and its package. An SPME method using PDMS fiber was developed to identify the odor causing compounds in a PVC-coated foil blister packaging [96]. The source of the odor was traced to be ethyl-2-mercaptoacetate, formed by the unanticipated reaction of a common residual solvent (ethanol) in a widely used pharmaceutical tablet excipient with low-level residual amounts of reactants or synthetic intermediates of PVC-resin thermal stabilizing agent. This compound was easily identified after extraction with SPME, while its concentration was below the detection limit of conventional headspace sampling. An on-line in-tube SPME method was developed for analyzing the contamination of endocrine disruptors in liquid medicines and intravenous injection solutions [97]. The quantified compounds included bisphenol A, alkylphenols and phthalates and they were separated and detected by high performance liquid chromatography with UV detection. The limits of quantification were 1–10 $ng\,mL^{-1}$. Phthalates were detected in the intravenous solutions contained in plastic containers, phthalates were also detected in syrup, lotion and eye drops, and DEHP easily leached from polyvinyl chloride tubing. Bisphenol A and alkylphenols were not detected in the studied solutions. Almost 50 different compounds were extracted by polar polyacrylate fiber from resin-based dental materials [98]. The main leachables included ethylene glycol dimethacrylate, iodobenzene, camphorquinone and *tert*-butyl-*p*-hydroxyanisole.

The solid phase microextraction method was developed to determine the release profile of a new flexible medical grade PVC/PCL-PC blend during aging in water at different temperatures [19, 36]. The low molecular weight compounds migrating from PVC/PCL-PC included degradation products of PCL-PC, different carboxylic acids originating from epoxidized soya bean oil and some other synthesis related compounds (phenol, acetophenone). SPME-GC-MS provided a rapid tool to identify and quantify compounds migrating from the PVC blend during aging. SPME-GC-MS results correlated well with weight loss results and surface analysis. In an unpublished work, the migration of low molecular weight compounds from the same PVC/PCL-PC material during steam sterilization for 20 minutes at 125 °C was studied. Low molecular weight compounds released during sterilization were extracted by CW/DVB solid-phase microextraction fiber and subsequently identified by GC-MS. The low molecular weight compounds included degradation products from PCL-PC and epoxidized soya bean oil and impurities remaining from synthesis, i.e., acetophenone, caprolactone, phenol, 6-hydroxyhexanoic acid and hexadecanoic acid.

7 Future Perspectives

SPME is a versatile extraction technique with great potential in polymer analysis and characterization. SPME has already been applied for the extraction of volatiles and semi-volatiles from polymeric materials including food packaging, medical and pharmaceutical materials, automotive materials and environmentally degradable polymers. It has also been shown that SPME is a valuable tool for the quality control of recycled materials, for extraction of polymer additives from different environmental samples such as water, sewage and sediment, for degradation and stability studies, for the identification of odor compounds and for control of rest monomers. In the future, it is expected that the total composition of polymeric materials will become even more of an important issue governed by strict regulations. Possible future applications of the SPME technique include rapid quality control of polymeric materials and products, reaction monitoring, screening tests for low molecular weight compounds and migration studies. Due to the rapid extraction and low impact on the sample, SPME techniques could be very valuable for studying degradation intermediates during oxidation, thermal degradation or biodegradation of polymers. On-fiber degradation studies could provide unique and reliable qualitative and quantitative information of photo-transformation of polymer additives and oligomeric model compounds. SPME could also be applied for studying the release profiles of different slow release systems for agriculture and biomedical applications. In vivo sampling techniques utilising SPME are currently being developed and could become valuable tools for evaluating the safety and function of biomedical materials.

References

1. Arthur CL, Pawliszyn J (1990) Anal Chem 62:2145
2. Pawliszyn J (1997) Solid Phase Microextracton – Theory and Applications. John Wiley & Sons, Inc., N.Y., USA
3. Pawliszyn J (ed) (1999) Applications of solid phase microextraction, RSC Chromatography monographs. The Royal Society of Chemistry, Cambridge, UK
4. Hakkarainen M, Albertsson A-C (2005) Biomacromolecules 6:775
5. Hakkarainen M, Burman L, Albertsson A-C (2006) In: Khemani K, Scholz C (eds) ACS Symp Series No 939 Degradable Polymers and Materials: Principles and Practice, Chapter 19, Amer Chemical Soc, Washington, USA, p 307
6. Hakkarainen M, Karlsson S, Albertsson A-C (1999) Polymer 41:2331
7. Albertsson A-C, Erlandsson B, Hakkarainen M, Karlsson S (1998) J Environ Polym Degrad 6:187
8. Potyrailo RA, Wroczynski RJ, Morris WG, Bradtke GR (2004) Polym Degrad Stab 83:375
9. Hakkarainen M, Albertsson A-C, Karlsson S (1997) J Environ Polym Degrad 5:67
10. Khabbaz F, Albertsson A-C, Karlsson S (1998) Polym Degrad Stab 61:329
11. Khabbaz F, Albertsson A-C, Karlsson S (1999) Polym Degrad Stab 63:127
12. Khabbaz F, Albertsson A-C (2000) Biomacromolecules 1:665
13. Hakkarainen M, Gröning M, Albertsson A-C (2003) J Appl Polym Sci 89:867
14. Gröning M, Hakkarainen M (2001) J Chromatogr A 932:1
15. Gröning M, Hakkarainen M (2002) J Appl Polym Sci 86:3396
16. Gröning M, Hakkarainen M (2004) J Chromatogr A 1052:151
17. Hakkarainen M (2007) J Biochem Biophys Methods 70:229
18. Gröning M, Hakkarainen M (2004) J Chromatogr A 1052:61
19. Hakkarainen M (2003) J Chromatogr A 1010:9
20. Gallet G, Erlandsson B, Albertsson A-C, Karlsson S (2002) Polym Degrad Stab 77:55
21. Chambers DM, Bazan JM, Ithaca JI (2001) Polymer Preprints 42:298
22. Ezquerro Ó, Pons B, Tena MT (2002) J Chromatogr A 963:381
23. Ezquerro Ó, Pons B, Tena MT (2003) J Chromatogr A 985:247
24. Ezquerro Ó, Pons B, Tena MT (2003) J Chromatogr A 999:155
25. Ezquerro Ó, Pons B, Tena MT (2003) J Chromatogr A 1020:189
26. Ortiz G, Tena MT (2006) J Chromatogr A 1101:32
27. Ezquerro Ó, Pons B, Tena MT (2003) J Chromatogr A 1008:123
28. Zhou S, Huang JX, Kinkel J, Stark W (2006) Bull Environ Contam Toxicol 76:301
29. Bortoluzzi JH, Pinheiro EA, Carasek E, Soldi V (2005) 89:33
30. Hall AD, Patel M (2006) Polym Degrad Stab 91:2532
31. Vance AL, Alviso CT, Harvey CA (2006) Polym Degrad Stab 91:1960
32. Rogalewicz R, Voelkel A (2005) Chemia Anal 50:991
33. Eriksson M, Swartling A, Dalhammar G, Faldt J, Borg-Karlsson AK (1998) Appl Microbiol Biotechnol 50:129
34. Dong RA, Liao PL (2001) J Chromatogr A 918:177

35. Gallet G, Lempiäinen R, Karlsson S (2001) Polym Degrad Stab 71:147
36. Hakkarainen M (2003) Polym Degrad Stab 80:451
37. Penton Z, SPME Varian Application Note, number 7.
38. Kusch P, Knupp G (2002) J Sep Sci 25:539
39. Kusch P, Knupp G (2004) J Polym Environ 12:83
40. Henry AC, McCarley RL, Das SS, Khan Malek GC (1999) J Electrochem Soc 146:2631
41. Penton Z, SPME Varian application note, number 17.
42. Charvet R, Cun C, Leroy P (2000) Analysis 28:980
43. Batlle R, Sanchez C, Nerin C (2001) J AOAC Int 84:431
44. Latini G (2000) Biol Neonate 78:269
45. Tickner JA, Schettler T, Guidotti T, McCally M, Rossi M (2001) Am J Ind Med 39:100
46. Hill SS, Shaw BR, Wu AHB (2001) Clin Chim Acta 304:1
47. Li X, Zeng Z, Chen Y, Xu Y (2004) Talanta 63:1013
48. Saito Y, Nakao Y (1999) Chromatography 20:124
49. Prokupková G, Holadová K, Poustka J, Hajšlová J (2002) Anal Chim Acta 457:211
50. Kelly MT, Larroque M (1999) J Chromatogr A 841:177
51. Penalver A, Pocurull E, Borrull F, Morcé MM (2000) J Chromatogr A 872:191
52. Saito J, Nakao Y, Imaizumi M, Morishima Y, Kiso Y, Jinno K (2002) Anal Bioanal Chem 373:81
53. Psillakis E, Kalogerakis N (2003) J Chromatogr A 999:145
54. Casajuana N, Lacorte S (2003) Chromatographia 57:649
55. Polo M, Llompart M, Garcia-Jares C, Cela R (2005) J Chromatogr A 1072:63
56. Luks-Betlej K, Popp P, Janoszka B, Paschke H (2001) J Chromatogr A 938:93
57. Möder M, Popp P, Pawliszyn J (1998) J Microcol Sep 10:225
58. Saba A, Pucci S, Raffaelli A, Salvadori P (1999) Rapid Commun Mass Spectrom 13:966
59. Frasch HF, Barbero AM (2005) Toxicol In Vitro 19:253
60. Hakkarainen M, Albertsson A-C, Karlsson S (2003) Int J Polymer Anal Charact 8:279
61. Isetun S, Nilsson U, Colmsjö A, Johansson R (2004) Anal Bioanal Chem 378:1847
62. Duffield P, Bourne D, Tan K, Garruto RM, Duncan MW (1994) J Anal Toxicol 18:361
63. Huppert N, Würtele M, Hahn HH (1998) Fresenius J Anal Chem 362:529
64. De Wit CA (2002) Chemosphere 46:583
65. Polo M, Gómez-Noya G, Quintana JB, Llompart M, García-Jares C, Cela R (2004) Anal Chem 76:1054
66. Polo M, Llompart M, Garcia-Jares C, Gomez-Noya G, Bollain M-H, Cela R (2006) J Chromatogr A 1124:11
67. Sánchez-Prado L, Llompart M, Lores M, Garcia-Jares C, Cela R (2005) J Chromatogr A 1071:85
68. Sánchez-Prado L, Gonsález-Barreiro C, Lores M, Llompart M, García-Jares C, Cela R (2005) Chemosphere 60:922
69. Sánchez-Prado L, Lores M, Llompart M, García-Jares C, Bayona J M, Cela R (2006) J Chromatogr A 1124:157
70. Salgado-Petinal C, Llompart M, García-Jares C, García-Chao M, Cela R (2006) J Chromatogr A 1124:139
71. Salgado-Petinal C, Garcia-Chao M, Llombart M, Garcia-Jares C, Cela R (2006) Anal Bioanal Chem 385:637
72. Alzaga R, Peña A, Bayona JM (2003) J Sep Sci 26:87
73. Walker V, Mills GA (2001) Clin Chim Acta 306:51
74. Del Olmo M, Zafra A, Suárez B, Gonzalez-Casado A, Taoufiki J, Vílchez JL (2005) J Chromatogr B 817:167
75. Kuruto-Niwa R, Terao Y, Nozawa R (2002) Environ Toxicol Pharmacol 12:27

76. Batlle R, Colmsjö A, Nilsson U (2001) Fresenius J Anal Chem 371:514
77. Burman L, Albertsson A-C, Höglund A (2005) J Chromatogr A 1080:107
78. Cruz SA, Zanin M, Nerin C, De Moraes MAB (2006) Food Additives Contaminants 23:100
79. Kontominas MG, Goulas AE, Badeka AV, Nerantzaki A (2006) Food Addit Contam 23:634
80. Strathmann S, Pastorelli S, Simoneau C (2005) Sens Actuators B 106:83
81. Cizkova H, Voldrich M, Dobias J (1998) Czech J Food Sci 16:81
82. Schwope AD, Till DE, Ehntholt DJ, Sidman KR, Whelan RH, Schwartz PS, Reid RC (1987) Food Chem Toxicol 25:317
83. Tombesi NB, Freije H (2002) J Chromatogr A 963:179
84. Suáres S, Sueiro RA, Garrido J (2001) Mutat Res Gen Toxicol Environ Mutagens 490:221
85. Nerín C, Philo MR, Salafranca J, Castle L (2002) J Chromatogr A 963:375
86. Salafranca J, Batlle R, Nerín C (1999) J Chromatogr A 864:137
87. Kataoka H, Ise M, Narimatsu S (2002) J Sep Sci 25:77
88. Chang C-M, Chou C-C, Lee M-R (2005) Anal Chim Acta 539:41
89. Feng Y-L, Zhu J, Sensenstein R (2005) Anal Chim Acta 538:41
90. Jiang GB (2001) J Hyg Res 30:1
91. De Santiago A, Aguilar-Santelises M (1999) Hum Exp Toxicol 18:619
92. Hoch M (2001) Appl Geochem 16:719
93. Ou QR, Whang CW (2006) Anal Bioanal Chem 386:376
94. Azenha M, Vasconcelos MT (2002) J Agric Food Chem 50:2713
95. Liu J-Y, Jiang G-B (2002) J Agric Food Chem 50:6683
96. Sides SL, Polowy KB, Thornquest AD Jr, Burinsky DJ (2001) J Pharm Biomed Anal 25:379
97. Mitani K, Narimatsu S, Izushi F, Kataoka H (2003) J Pharm Biomed Anal 32:469
98. Rogalewicz R, Voelkel A, Kownacki I (2006) J Environ Monit 8:377

Adv Polym Sci (2008) 211: 51–84
DOI 10.1007/12_2007_117

Published online: 9 October 2007

Quantitative Determination of Volatiles in Polymers and Quality Control of Recycled Materials by Static Headspace Techniques

Mikael Gröning · Minna Hakkarainen · Ann-Christine Albertsson (✉)

Department of Fibre and Polymer Technology,
School of Chemical Science and Engineering,
Royal Institute of Technology (KTH), 10044 Stockholm, Sweden
aila@polymer.kth.se

Abstract A presentation is given of headspace (HS) extraction and headspace solid-phase microextraction (HS-SPME) techniques and their combination with multiple headspace (MHS) extraction to enable quantitative determination of volatiles in solid polymer matrixes. As an example, the development of HS, HS-SPME, and MHS-SPME methods for extraction of volatiles from thermo-oxidized and/or recycled polyamide 6.6 is reviewed with special focus on the problems encountered when extracting analytes from solid-sample matrixes including excessively long equilibrium times and adsorption of analytes to the sample matrix. Examples are also given of the application of HS-SPME in quality control of recycled materials, in durability assessment of polymeric materials and in degradation studies.

Keywords Degradation · Multiple headspace extraction · Quality control · Recycling · Solid-phase microextraction

Abbreviations

ASTM	American Society for Testing Materials
BTEX	Benzene, toluene, ethylbenzene, xylene isomers
CAR	Carboxen
CW	Carbowax
DSC	Differential scanning calorimetry
DVB	Divinylbenzene
DMF	Dimethyl formamide
FTIR	Fourier transform infrared spectroscopy
GC	Gas chromatography
HS	Headspace
IR	Infrared spectroscopy
LC	Liquid chromatography
MAE	Microwave-assisted extraction
MHE	Multiple headspace extraction
MHS	Multiple headspace
NMR	Nuclear magnetic resonance spectroscopy
PA	Polyacrylate
PA 6	Polyamide 6
PDMS	Polydimethylsiloxane
PP	Polypropylene
RSD	Relative standard deviation
SDME	Single-drop microextraction
SPME	Solid-phase microextraction
SFE	Supercritical fluid extraction
TGA	Thermogravimetric analysis
TVOC	Total volatile organic content
USE	Ultrasound extraction
UV	Ultraviolet spectroscopy

1 Introduction

Plastic products contain residual monomers, by-products from polymerization, additives and degradation products. The low-molecular-weight compounds will in time migrate from the polymers into the surrounding environment. Thus, the knowledge of the amounts and identities of these compounds is important both to producers and consumers of plastics. For both scientific and engineering purposes, it would be valuable to be able to assess the remaining life-time of a polymer by measuring the content of degradation products. Extraction of low-molecular-weight compounds simply by heating the polymer, i.e. without using chemical solvents, should for a number of reasons be the preferred choice:

- As chemical solvents are not required, or are used only in small amounts, the environmental load of the analysis is reduced
- The risk of losing volatile analytes is reduced as the sample is generally heated inside a closed system
- No solvent peak that can mask early eluting volatiles appears in the chromatogram

Headspace extraction relies on the heating of the sample to a temperature where a sufficiently high amount of the analyte is volatilized to reach a concentration of analyte in the gaseous phase above the sample to meet the sensitivity limits of the employed technique. For some samples, heating is not possible due to low thermal stability of the analyte or the polymer, and for such systems, the analyst is limited to leaching the analytes from the polymers using chemical solvents, followed by chromatographic separation and analysis. As the solid sample is put in contact with the solvent, without dissolving the polymer, the analytes will leach from the polymer to the solvent and the analytes can subsequently be identified using, e.g. gas chromatography (GC) or liquid chromatography (LC). In the most common leaching technique, Soxhlet, the migration of analyte from polymer to solvent is aided by high extraction temperature [1]. Other leaching techniques use microwaves, microwave-assisted extraction (MAE), ultrasound (USE) or supercritical fluids (SFE) to facilitate leaching [2–6]. Solvent extractions are best suited for analytes of medium to high molecular weight, such as additives and oligomers, as volatiles are easily lost during sample handling. Generally, cleanup and concentration is required prior to chromatographic analysis.

Volatiles in polymers can also be analyzed by other techniques such as thermogravimetric analysis (TGA) or different spectroscopic techniques. TGA is the most general method for measuring the volatile content of a polymer. TGA monitors the weight loss of a sample that is heated to a sufficiently high temperature to volatilize the low-molecular-weight compounds present in the polymer. Such a measurement only yields the total volatile content and no identification of the volatiles is obtained. Additives and volatiles can also be analyzed directly in solid polymers using ultraviolet (UV), infrared (IR) or nuclear magnetic resonance (NMR) spectroscopy. Spectroscopic techniques are more selective than TGA for analysis of solids and provide structural information of the analyte, but suffer from poor selectivity and sensitivity, narrow linear dynamic range and difficulties in quantitation due to the unavailability of solid standards [7]. Spectroscopic techniques are primarily employed for the analysis of medium to high mass compounds, which, due to slow migration, are difficult to extract from the polymers.

2 Principles of Static Headspace Extraction

Headspace extraction means sampling from the headspace, i.e. the gaseous phase, above the solid sample that is heated in a closed vessel [8]. Headspace extraction does not use any solvents and thus involves no interfering solvent peak in the chromatogram. The principle of headspace extraction is that volatile compounds in any liquid or solid matrix (both referred to as the condensed state) will also be present in the gaseous phase above the condensed sample. The analytes are distributed between the gaseous and condensed phases according to a thermodynamically controlled equilibrium [9]. The actual mass distribution is controlled by the solubility of the analyte in the condensed phase and compounds with high solubility will, thus, have a high concentration in the condensed phase relative to the gaseous phase. On the other hand, for analytes with low solubility in the condensed phase, the concentration in the condensed phase will be close to that in the gaseous phase, or even smaller. In headspace extraction of volatiles in polymers, the mass distribution is mostly determined by the volatility of the analyte. When a headspace extraction method is developed the sensitivity can be optimized by changing either the temperature or the ratio between sample and headspace volume in the sample vial. As a general guideline, it can be stated that high headspace sensitivity for analytes with high solubility in the condensed phase is obtained using high equilibrium temperature. On the other hand, high headspace sensitivity for analytes with low solubility in the condensed phase is generally achieved using a large sample volume, i.e. a low headspace-to-sample volume ratio.

According to the traditional headspace extraction technique, the sampling from the headspace is done either in a static or continuous mode [10]. The static mode uses a heated syringe to withdraw an aliquot of the headspace and for high sensitivity much attention has to be paid to optimizing the equilibrium temperature and sample volume. In the continuous mode, the sample is continuously flushed with a stream of inert gas that, in time, theoretically, removes all the analyte from the sample. Although dynamic headspace extraction generally gives a complete recovery of analytes when used for liquid samples, the extraction is seldom exhaustive when applied to polymer samples [11–13]. Continuous extraction requires less control of extraction parameters; the volume of the extracting gas and the extraction temperature being the most important parameters, but it involves other significant drawbacks [14]. For example, the high volume of the extracting gas used dilutes the analytes and makes it necessary to use a focusing device, such as an adsorption or cryofocusing trap, prior to chromatographic analysis [15]. The first report on the use of headspace extraction as an introduction technique for chromatographic analysis was presented in 1958 [16]. It was followed by a rapid development of the headspace technique and the first system for automated chromatographic analysis of headspace samples was introduced to the market in the late 1960s. Reports on the application of headspace extraction followed by gas chromatographic separation in polymer analysis such as the determination of ethylene oxide in sterilized materials [17] and residual styrene in poly(styrene-co-α-methyl-styrene) [18] appeared in the literature in the early 1970s. Today both static and dynamic headspace extractions are commonly used for analyzing volatiles in polymers. The American Society for Testing Materials (ASTM International) also describes a general procedure for the qualitative analysis of volatiles in polymers by static HS [19].

2.1 Headspace Solid-Phase Microextraction

Solid-phase microextraction (SPME) is a rapid, sensitive and solvent-free technique for extraction of low-molecular-weight compounds. The solid-phase microextraction technique was developed by Arthur and Pawliszyn and first presented in 1990 [20]. The technique has gained wide popularity in various areas. However, it has not yet been widely used in the analysis of low-molecular-weight compounds in polymers. In SPME, an approximately 1-cm-long fused-silica fiber coated with a polymeric phase is used for extraction. The fiber is mounted for protection in a syringe-like device. Although the SPME-technology was originally developed for sampling from aqueous samples by immersing the fiber into the sample (direct extraction), its applicability for headspace extraction from condensed samples was shown by Zhang and Pawliszyn in 1993 [21]. During the extraction, the analytes are absorbed or adsorbed by the extracting phase, depending on the nature of the coating [22]. After the completed extraction, the analytes are desorbed in the injector of a gas or liquid chromatograph for further separation and identification [23].

In the direct extraction mode, the SPME partitioning resembles liquid–liquid extraction and, besides the type of fiber coating used, parameters important for optimal recovery are. e.g. agitation technique and pH. For headspace SPME, the most important parameters are the vial volume, the headspace-to-sample volume ratio and the equilibrium temperature, i.e. similar parameters as in traditional HS [24, 25]. Although the theoretical treatment of SPME relies on extraction under equilibrium conditions, it is not necessary to establish full equilibrium to perform quantitative analysis. It was shown by Ai that quantitative data can be obtained from both two- and three-phase systems under non-equilibrium conditions [26–28]. This expands the use of SPME significantly as many types of samples require an exceedingly long time to establish equilibrium. However, when working under non-equilibrium conditions, consistent timing is very important as small deviations in extraction time may result in large differences in the amount of analyte extracted.

The selectivity of the extraction and the amount of analyte extracted is determined by the partition coefficient (distribution ratio) of the analyte between the sample matrix and the fiber coating (direct SPME) or between sample matrix, headspace, and the coating material (HS-SPME). One great advantage of SPME compared to traditional static headspace extraction originates from the partitioning coefficient of the SPME extracting phase. When an appropriate fiber coating is used, the analyte will favor the coating phase rather than the sample matrix, which results in an enrichment of the analyte in the fiber. This gives a higher sensitivity for SPME compared to static headspace sampling, which has been demonstrated by extracting degradation products from polyethylene [29] and odor-causing volatile compounds in packaging materials [30]. In addition, HS-SPME has been shown to be more sensitive than both static and dynamic HS for the extraction of flavor compounds in milk [31] and for the extraction of volatiles in a dynamic system [32]. There are two standard procedures issued by ASTM International for the analysis of volatiles and semi-volatiles in water by SPME [33, 34].

Proper selection of the fiber coating is essential for obtaining good sensitivity by SMPE. In general, polar analytes are best extracted by polar coatings such as polyacrylate or Carbowax/divinylbenzene (CW/DVB), whereas non-polar analytes are best extracted by a non-polar coating such as polydimethylsiloxane (PDMS). There are several different coatings commercially available that together make possible the extraction of a wide range of analytes.

3 Quantitative Techniques in Headspace Analysis of Volatiles in Polymers

In most cases, the analyte is only partially extracted from the sample during headspace extraction. In an ASTM procedure, the partitioning of vinyl chloride monomer between poly(vinyl chloride) and its headspace at an equilibrium temperature of 90 °C is reported, which allows external calibration in the ppm range by headspace extraction [35]. However, if the recovery of the extraction is unknown, quantitation is not possible. To determine the recovery of an extraction, standards prepared from the same matrix as the sample and containing known amounts of analyte are extracted. Such standards are easily prepared if the matrix is gaseous or a liquid containing only a few components. After measuring the recovery of flavor compounds by headspace SPME from spiked wine and beer, external calibration was used for quantitation [36–38]. For liquid and semi-solid samples containing many interfering compounds, it may not be possible to produce standards that accurately reflect the recovery of the analyte from the true samples, due to adsorption of the analyte. During the analysis of volatiles in solid polymers, the main obstacle to quantitation is the problem of mixing volatiles with solid polymers, as the volatiles will be lost at the high temperatures required to melt the polymers. This often limits the extraction of low-molecular-weight compounds from polymers to qualitative rather than quantitative measurements.

In industry, the content of volatiles in plastics and other solid samples is routinely estimated in order to quality control the products. Due to the previously mentioned obstacles for performing quantitative analysis of volatiles in solid samples, a simplified HS-GC method is generally used [39, 40]. The total volatile organic content (TVOC) is estimated by comparing the sum of all peak areas in the resulting chromatogram to a calibration with acetone or toluene as a standard and the TVOC is reported in μg C/g. The measurement does not give the absolute volatile content, as it is performed under non-equilibrium conditions, and no compensation is made for sample volume, matrix effects or differences in volatility between analytes and the standard used. However, the values given are associated with threshold values that restrict the use of the materials in certain applications.

Only in rare cases are volatile analytes exhaustively extracted from polymeric samples. In such cases, the amount of analyte in the sample is easily calculated by simple external calibration, i.e. from a correlation between the detector response and the amount of analyte introduced to the analytical instrument. Examples of exhaustive extractions are headspace extractions of methyl methacrylate from polymer latex [41] and acetaldehyde from polyethylene terephthalate [42]. In some cases, the standard addition technique provides a solution to the problem of preparing spiked standards. Standard addition involves the addition of known quantities of analyte to the sample prior to extraction. The amount of analyte in the original sample is obtained by comparing the peak area of the analyte in the original sample to the peak area of the analyte in the sample with a known amount of added analyte. ASTM describes procedures to quantitate residual vinyl chloride monomer in vinyl chloride homo- and co-polymers and acrylonitrile monomer in styrene-acrylonitrile co-polymers, respectively, by standard addition and HS-GC. As described in the procedures, the polymers are dissolved in suitable solvents prior to the addition of known amounts of monomer. These methods allow quantitation in the ppm to ppb range [43, 44]. A well-known problem with the technique of standard addition is that when working with strongly adsorptive matrices such as soil, native and externally added analytes may not be adsorbed equally strong by the sample. This was observed by Eriksson et al., for example, who compared the quantitation of aromatic hydrocarbons in soil by HS-SPME to liquid extraction [45]. Standard addition has been used successfully together with HS-SPME to quantitate volatile flavor additives in tobacco [46] and environmental pollutants in fish tissue [47].

3.1 Multiple Headspace Extraction

In 1977 Kolb and Pospisil proposed a method for the quantitative analysis of volatiles in solid samples [48] by using headspace extraction and gas chromatographic detection. The method, termed discontinuous gas extraction, is based on stepwise gas extraction, followed by a subsequent analysis of the extracted volatiles. The method theoretically calculates the total amount of analyte in a solid sample after a few successive extractions and makes the quantitation of volatile analytes in solid matrices possible. The proposed method was validated by measuring the styrene content in polystyrene by discontinuous gas extraction and by a procedure proposed by Rohrschneider in which the polystyrene is dissolved in dimethyl formamide (DMF) [49]. The two methods were in good agreement, which supported the validity of the discontinuous gas extraction. Kolb and Pospisil later elaborated the theoretical treatment of discontinuous gas extraction and in 1981 the method was re-named as multiple headspace extraction (MHE) [50].

The principle of the MHE procedure is based on stepwise gas extractions at equal time intervals. When a portion of the headspace is removed in the first extraction, the equilibrium between the analyte in the condensed phase and in the headspace is disturbed. As the sample is allowed to re-equilibrate, the analyte will migrate from the condensed phase into the headspace until the same ratio between the concentrations in the two phases as in the first extraction is obtained. The concentrations in the two phases will now be smaller than they were during the first extraction. A second analysis will thus result in a smaller peak and by continuing this procedure it is possible to strip off all the volatiles from the sample. If carried out ad infinitum, the various peak areas are summed up to get the total peak area, which corresponds to the total amount of the analyte in the sample. The influence of sample matrix is thus eliminated by exhaustive extraction. As the MHE procedure strictly follows a logarithmic function it is not required that the extractions are carried out until all the analyte is removed from the condensed sample. Instead, the logarithms of the various area values from the consecutive analyses are plotted versus the number of analyses in a linear scale and the total area value is obtained by regression calculation from the areas obtained in only a few extraction steps [51]. Although the MHE method was received with some scepticism [52, 53] when first proposed, later, through a thorough theoretical treatment by Kolb and Ettre [54], it was shown to be valid. The combination of multiple headspace technique with solid-phase microextraction (MHS-SPME) was recently reviewed in two papers [55, 56]. Multiple headspace extraction was also recently combined with single-drop microextraction (SDME) to enable the quantitative determination of styrene in polystyrene [57].

3.2
Limitations of Multiple Headspace Extraction

Problems may arise in quantitation by MHE when both the analyte and the sample matrix are polar. In this case, the system under investigation is referred to as an adsorption system, due to the adsorption of analyte by the sample matrix [58]. Systems that are free from analyte adsorption are designated partitioning systems. Only solid samples can form adsorption systems, whereas liquid samples always form partitioning systems. The presence of adsorptive forces manifests itself in the graph of logarithms of peak areas versus the number of successive extractions. If the analyzed system is an adsorption system, there will be no exponential decrease in peak area throughout the successive extractions. Thus, in the regression plot there will be a non-linear relationship between the logarithms of peak areas versus the number of extraction steps. If adsorption is identified, the system can be transferred from an adsorption system into a partitioning system by adding a compound with higher affinity towards the adsorbing sites in the matrix than the analyte of interest. Such a compound is referred to as a displacer or modifier. Adsorption of analytes is a well-known problem when extracting environmental pollutants from soil. Cyclohexanone could be quantitatively analyzed in soil by multiple headspace extraction using water as a displacer [59]. The addition of water increased the recovery dramatically, from 4% to 99.4%, as determined from spiked samples. Also, BTEX (benzene, toluene, ethylbenzene and xylene isomers) in soil could be analyzed quantitatively using multiple headspace SPME [60] when water was added to the sample. Measurements of the BTEX content in a certified reference sample showed the excellent accuracy of the MHS-SPME method. Water is a good displacer and has also been used in the headspace extraction of aldehydes from cardboard [61]. When static headspace extraction was used, the extracted amount of hexanal increased six times, clearly showing the higher affinity of the polar adsorption sites of the matrix towards water compared to hexanal [62]. The addition of water also allowed the quantitation of pentanal in cardboard using multiple headspace extraction [63].

4
Development of HS-SPME Methods for Extraction of Volatiles from Polyamide 6.6

Many factors are important for successful extraction and particularly the quantitative determination of volatiles by HS-SPME. As an example, this chapter reviews the method development for HS-SPME of volatiles from polar polyamide 6.6 matrix including the effect of fiber material, extraction time, incubation time, extraction temperature and quantitative determination by

MHS-SPME [64–67]. The additional problems caused by polar matrix are also discussed.

4.1 Fiber Coating

Proper selection of fiber coating is essential for obtaining good sensitivity. Initially, three different coating materials, polydimethylsiloxane/divinylbenzene (PDMS/DVB), carbowax/divinylbenzene (CW/DVB) and polyacrylate (PA), were evaluated for the extraction and screening of polyamide 6.6 thermo-oxidation products [64]. These fibers were chosen due to their suitability for extraction of volatile and semi-volatile polar compounds as it is expected that mainly polar degradation products are formed during the thermo-oxidation of polyamide 6.6. To test the performance of the selected fibers, degradation products formed during 100 h of thermo-oxidation from polyamide 6.6 strips were extracted by the three fibers. The extraction time and temperature were 30 min and 80 °C, respectively. The PDMS/DVB fiber extracted more compounds and in larger quantities than the other fibers. 2-cyclopentyl-cyclopentanone was one of the most abundant compounds extracted from polyamide 6.6 after the thermo-oxidation and it was selected as a target analyte for quantitation studies. Previous studies have also identified 2-cyclopentyl-cyclopentanone as the most abundant low-molecular-weight compound in polyamide 6.6 [68–71].

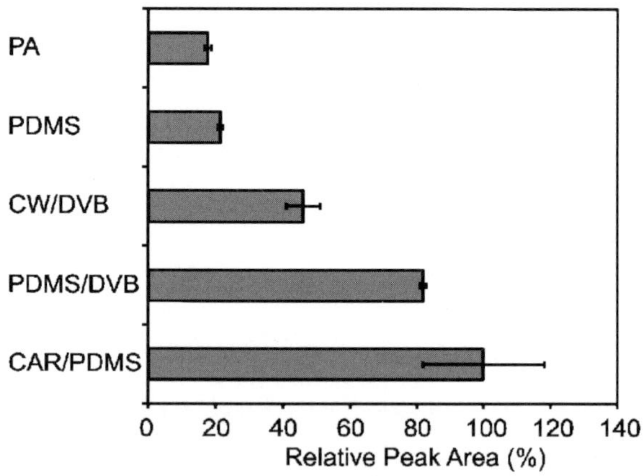

Fig. 1 Relative peak areas and standard deviations after triplicate headspace extractions of 100-ng 2-cyclopentyl-cyclopentanone standard. Extraction time and temperature was 30 min at 80 °C. Fiber material PA = polyacrylate; PDMS = polydimethylsiloxane; CW/DVB = carbowax/divinylbenzene; PDMS/DVB = polydimethylsiloxane/divinylbenzene, CAR/PDMS = carboxen/polydimethylsiloxane. Reprinted from [66] with permission of Elsevier. © Elsevier (2004)

Five different fiber materials PDMS/DVB, CW/DVB, PA, PDMS and carboxen/polydimethylsiloxane (CAR/PDMS) were evaluated for the extraction of 2-cyclopentyl-cyclopentanone [66]. Figure 1 shows the averages of relative peak areas after triplicate extractions of 2-cyclopentyl-cyclopentanone with the different fibers. The CAR/PDMS-fiber showed the highest recovery, but it also showed the lowest repeatability, with a relative standard deviation (RSD) of 18%. In addition, the 2-cyclopentyl-cyclopentanone peak showed excessive tailing when extracted with the CAR/PDMS-fiber. The peak symmetry could not be improved by increasing the injector temperature as tailing was still observed at injector temperatures of 270 °C and 300 °C. The PDMS/DVB fiber showed excellent RSD of 3% and the second best recovery. It also gave good peak symmetry even at an injector temperature of 220 °C. The PDMS/DVB fiber was thus chosen for the quantitation studies.

4.2 Linear Range of PDMS/DVB Fiber

Quantitative or semi-quantitative determination of analytes by SPME requires working within the linear dynamic range of the SPME fiber. If the linear dynamic range is exceeded, the extracted amount of analyte will not reflect the amount of analyte in the sample. Figure 2 shows the normalized peak areas for the headspace extractions of different amounts of powdered polyamide 6.6 [67]. The extraction time and temperature were 45 min and 80 °C. Under the given conditions, the dynamic range of the PDMS/DVB fiber was linear if the polyamide sample size was between 1 and 100 mg. For the

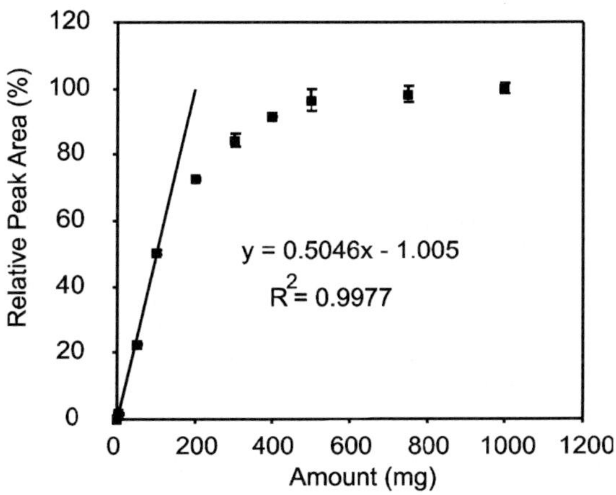

Fig. 2 The 45-min headspace extractions of 2-cyclopentyl-cyclopentanone from 1 to 1000 mg polyamide 6.6 at 80 °C using a PDMS/DVB fiber. The dynamic range was linear when the sample size was between 1 and 100 mg. Reprinted from [67] with permission of Elsevier. © Elsevier (2004)

quantitation studies, it was decided to use 75 mg of sample for each extraction of 2-cyclopentyl-cyclopentanone from polyamide 6.6 at 80 °C, as this gave a clear peak in the chromatograms and was within the linear dynamic range of the fiber.

4.3 Establishment of Equilibrium

Establishment of equilibrium is not required for quantitative determination of analytes by SPME. However, consistent timing is highly important if the extraction is stopped before the equilibrium is attained, as small deviations in extraction time may generate large differences in the extracted amount. When extractions are carried out in headspace mode, the partitioning between the analyte in the condensed phase and its headspace is generally slower than the partitioning between the analyte in the headspace and the fiber coating. The time required for reaching equilibrium between 2-cyclopentyl-cyclopentanone in powdered polyamide 6.6 samples and in the headspace was studied by incubating 25 mg of powdered polyamide 6.6 at 80 °C for up to 44 h [66]. HS-SPME was performed after different incubation times. The extraction time in all cases was 30 min. The extraction profile given in Fig. 3 clearly shows that equilibrium is reached first after 12 h of incubation at 80 °C.

The long time required for reaching the equilibrium is most likely caused by strong hydrogen bonding between the analyte and the polymer matrix. The relatively high degree of crystallinity (approximately 50%) in polyamide 6.6 may also contribute to the slow diffusion rate. When 2-cyclo-

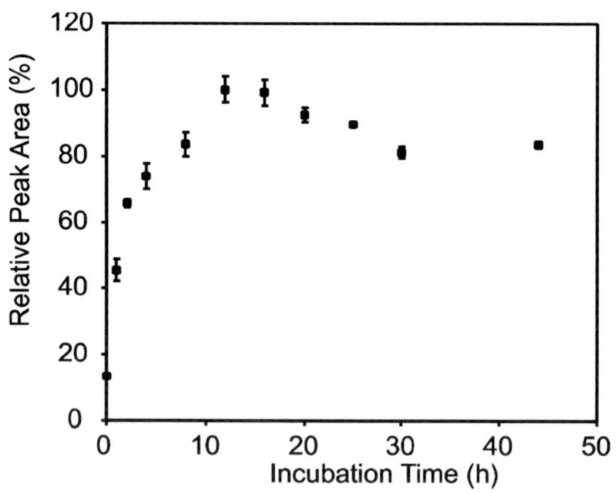

Fig. 3 Recovery profile after extraction from 25 mg powdered Zytel. Extraction time was 30 min at 80 °C. The samples were incubated at 80 °C for 0–44 h prior to extraction. Reprinted from [66] with permission of Elsevier. © Elsevier (2004)

pentyl-cyclopentanone was extracted in the absence of polyamide matrix at 80 °C equilibrium and optimal recovery was reached already after 20 min of extraction [66]. The recovery decreased in both cases upon prolonged heating. This drop in recovery could be due to the warming up of the SPME fiber, which is known to reduce the amount of analyte extracted. The standard deviation was rather large for the 20-min extraction but it decreased at extraction times over 30 min. The large standard deviation could also be explained by the changing fiber temperature during the first 30 min of extraction. The drop in recovery was not as large when extracting from a polymer matrix, as when extracting free 2-cyclopentyl-cyclopentanone. This is explained by the continuous migration of analyte from the sample matrix during the extraction. If true equilibrium has not been established, additional analyte will migrate from the sample during extraction, thus counteracting the drop in extraction efficiency due to heating of the SPME fiber, i.e. the solid sample acts as a reservoir of analyte.

In addition to the equilibrium reached after 12 h of incubation, two equilibrium-like states were observed during the first 60 min of extraction [66]. It was proposed that the equilibrium-like shape of the extraction profile, found during the first 30 min of extraction, represents the equilibrium between the readily available "free" 2-cyclopentyl-cyclopentanone on the surface of the polyamide 6.6 powder, the headspace and the SPME-fiber. The second equilibrium-like state probably represents equilibrium for the 2-cyclopentyl-cyclopentanone originally present inside the polyamide 6.6 powder, at a close distance to the outer surface. The equilibrium-like shape of the recovery profile may lead to an erroneous selection of extraction conditions, which in turn would lead to erroneous quantitation. The long equilibrium time reflects the need for rapid methods to estimate the volatile content under non-equilibrium conditions.

4.4 Quantitation of 2-cyclopentyl-cyclopentanone by MAE

A method for extraction of 2-cyclopentyl-cyclopentanone from polyamide 6.6 by MAE was developed to quantitate the amount of 2-cyclopentyl-cyclopentanone in the polyamide 6.6 samples [66] and to validate a MHS-SPME method for extraction of 2-cyclopentyl-cyclopentanone [67]. The method was optimized with respect to the type of solvent, extraction temperature, extraction time and sample-to-solvent ratio. Chloroform and methanol were evaluated as extracting solvents. After extraction at 90 °C for 30 min the highest recovery was achieved using methanol as a solvent. This is attributed to the better compatibility between polar polyamide 6.6 and polar methanol, which gives good swelling of the polyamide 6.6 matrix and more effective extraction of analyte. The effect of extraction time on the recovery of 2-cyclopentyl-cyclopentanone was studied by extracting 1.0 g of polyamide 6.6 in 10 ml methanol for different times varying from 5 to 60 min. 2-cyclopentyl-cyclopentanone was quite rapidly extracted by methanol. The maximum recovery was achieved after 45 min, but already after 1 min of extraction approximately 75% of the maximum recovery was achieved. Most of the 2-cyclopentyl-cyclopentanone in polyamide 6.6, thus, migrates into the extraction solvent already during the heating of the extraction vessel. To ensure that the high extraction temperature and long extraction time does not cause any degradation of 2-cyclopentyl-cyclopentanone, 2-cyclopentyl-cyclopentanone was heated in methanol at 90 °C for 45 min. GC-MS analysis showed that the amount of 2-cyclopentyl-cyclopentanone in the solution was the same before and after the heating and hence no analyte degradation took place. The recovery of the developed MAE method could not be determined due to the unavailability of standard polyamide 6.6 containing known amount of 2-cyclopentyl-cyclopentanone. However, MAE generally gives rather high recoveries, and the amounts measured by MAE were used as approximations of the total amount of 2-cyclopentyl-cyclopentanone in five different polyamide 6.6 samples to validate a developed MHS-SPME method [67].

4.5 Quantitation of 2-cyclopentyl-cyclopentanone by MHS-SPME

A prerequisite in MHS-SPME is that the area ratio of consecutive peaks is constant throughout the concentration range. In practice, the constant area ratio is checked by observing the spread of individual data points from a linear regression line in the plot of logarithms of the peak areas obtained in consecutive extractions versus the number of extractions. In MHE, the linear regression line should have a correlation coefficient of at least 0.998 to show that the matrix effects are truly absent. However, the repeatability of SPME is lower than the repeatability of conventional HS, and somewhat lower correlation coefficients can be accepted. For MHS-SPME the correlation coefficient should be examined together with the shape of the extraction plot and a clear linear trend should be observed in the extraction plot to ensure the absence of matrix effects.

The effect of sample amount, extraction temperature, incubation time and addition of a displacer on the measured amount of 2-cyclopentyl-cyclopentanone in polyamide 6.6 and the characteristics of the corresponding linear regression lines are shown in Table 1 [67]. Changing the sample amount from 50 to 75 mg did not significantly influence the measured amount of 2-cyclopentyl-cyclopentanone. The extraction temperature, however, strongly influenced the measured amount. After MHS-SPME at 50 °C, the measured concentration of 2-cyclopentyl-cyclopentanone was 0.47 μg/g. At 80 °C and 120 °C, the measured concentrations were 3.90 μg/g and 96.07 μg/g, respectively. The very low correlation coefficient of 0.686 obtained at 120 °C immediately indicates that the measurement is invalid. At 50 °C and

Table 1 The amount of 2-cyclopentyl-cyclopentanone in Zytel polyamide 6.6 measured by MHS-SPME under various conditions. Reprinted from [67] with permission of Elsevier. © Elsevier (2004)

$n = 3$	Amount (μg/g) Mean ± std	Equation slope	Intercept	Correlation $R^2 \pm$ std
Amount sample (80 °C)				
50 mg	3.97 ± 0.08	- 0.29 ± 0.01	17.58 ± 0.07	0.995 ± 0.002
75 mg	3.90 ± 0.53	- 0.21 ± 0.03	17.36 ± 0.02	0.975 ± 0.019
Temperature				
50 °C, 75 mg	0.47 ± 0.00	- 0.39 ± 0.04	16.03 ± 0.09	0.936 ± 0.029
80 °C, 75 mg	3.90 ± 0.53	- 0.21 ± 0.03	17.36 ± 0.02	0.975 ± 0.019
120 °C, 25 mg	96.07 ± 17.57	- 0.04 ± 0.01	15.22 ± 0.04	0.686 ± 0.020
Incubation (80 °C)				
0 h, 75 mg	3.90 ± 0.53	- 0.21 ± 0.03	17.36 ± 0.02	0.975 ± 0.019
25 h, 10 mg	2.08 ± 0.15	- 0.81 ± 0.10	15.05 ± 0.14	0.968 ± 0.025
With modifier (90 °C)				
30 mg	15.61 ± 1.22	- 0.20 ± 0.03	16.63 ± 0.28	0.991 ± 0.021

80 °C, the correlation coefficients were 0.936 and 0.975, respectively. However, a close examination of the MHS-SPME extraction plots, shown in Fig. 4, show that both measurements are invalid [67].

Figure 4a shows the extraction plot of MHS-SPME at 50 °C [67]. The extraction plot is approximately linear from the second to the fifth extraction. However, it flattens out between the fifth and the sixth extractions. In addition, the difference in the relative peak area between the first and second extraction is larger than the difference in the relative peak areas between the following extractions. This reflects the slow migration of analyte from the sample at the low temperature used for the extraction. In the first extraction, the readily available analyte is extracted from the sample. In the second extraction, the analyte has had limited time to migrate from inside of the polyamide 6.6 powder to replace the analyte removed in the first extraction. Between the fifth and sixth extraction the extraction plot flattens out, which is concluded to be due to the adsorption of the analyte to the polar polyamide 6.6 powder. In the extraction plot, the adsorption is observed only after several extractions when the amount of "free" analyte in relation to the hydrogen-bonded analyte has decreased.

The shape of the extraction plot for the extraction at 80 °C is shown in Fig. 4b [67]. The plot is now linear from the first to the fifth extraction, but again it flattens out between the fifth and the sixth extraction, showing adsorption of analyte by the matrix even at 80 °C. Linearity is, however,

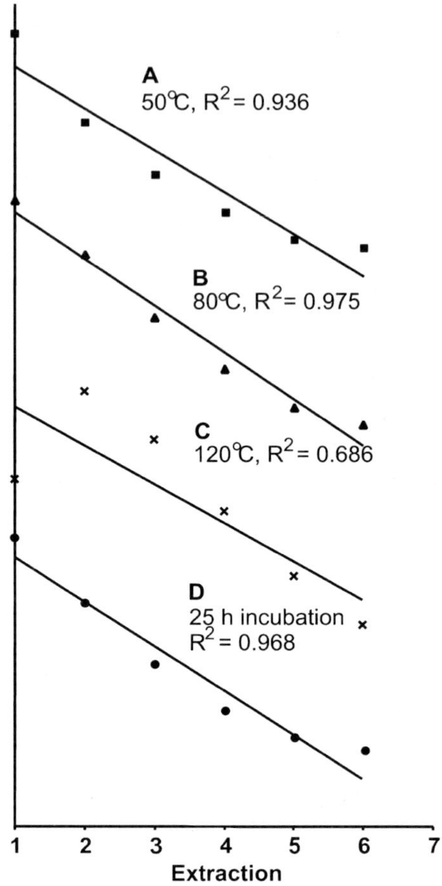

Fig. 4 The multiple headspace extraction plots of six consecutive 45-min extractions from: (**A**) 75 mg Zytel at 50 °C; (**B**) 75 mg Zytel at 80 °C; (**C**) 75 mg Zytel at 120 °C and (**D**) 10 mg Zytel at 80 °C. The 10-mg samples were incubated for 25 h at 80 °C prior to extraction. Reprinted from [67] with permission of Elsevier. © Elsevier (2004)

obtained at the beginning of the plot due to the faster migration rate at the higher temperature. This also results in higher correlation coefficient. At 80 °C, the 2-cyclopentyl-cyclopentanone concentration was measured to be 3.90 μg/g. This considerably higher value also reflects the faster migration rate at the higher temperature. However, due to the adsorption of the analyte, shown by the flattening of the extraction plot, this value is also considered erroneous. At 120 °C the faster migration rate released more analyte into the headspace than could be removed by the SPME-fiber in the first extraction, giving an extraction plot with a positive slope between the first and the second extraction (Fig. 4c). A small tendency to adsorption of analyte is still noticed as the extraction plot flattens out slightly between the fifth and sixth extractions. The concentration measured at 120 °C was 96.07 μg/g, but again the very poor average correlation coefficient of 0.686, together with the non-linear extraction plot, indicates erroneous results.

To overcome the problem of slow migration rate from the solid matrix, powdered polymer samples were incubated for 25 h at 80 °C prior to the 45 min extraction at 80 °C. The resulting extraction plot is shown in Fig. 4d [67]. The plot shows rather good linearity during the first four extractions, however the analyte is still adsorbed by the matrix as the extraction plot flattens out after the fourth extraction. The concentration of 2-cyclopentyl-cyclopentanone in Zytel was measured to be 2.08 μg/g when samples were incubated for 25 h at 80 °C prior to the extraction. Although the correlation coefficient of 0.968 is rather good, the shape of the extraction plot indicates that the measurement was still not valid. Adsorption of the analyte to the matrix was verified by multiple additions of 2-cyclopentyl-cyclopentanone to five different polyamide 6.6 samples followed by HS-SPME. To each sample, six different amounts in the range of 0 to 48 ng were added. The samples were stored for 24 h at room temperature to allow equilibrium between the solid sample and the added analyte prior to extraction for 45 min at 80 °C. The amount of 2-cyclopentyl-cyclopentanone in the samples was calculated by extrapolating the linear regression line to the intercept with the x-axis, which corresponds to the concentration of 2-cyclopentyl-cyclopentanone in the samples. However, the calculated amounts were only 20–40% of the amounts determined by MAE. The lower amounts measured results from the slow migration and adsorption of the analyte, which makes the peak areas corresponding to the sample without any additional analyte too small.

4.5.1 The Use of Water as a Displacer

Strong hydrogen bonding between 2-cyclopentyl-cyclopentanone and the polar polyamide 6.6 was concluded to be responsible for the previously discussed adsorption of analyte by the matrix [67]. Although the slow migration rate may be overcome by incubating the samples prior to extraction, the problem of adsorption of analyte by the matrix still remains. The adsorption of analyte may be suppressed by adding a compound with higher affinity than the analyte to the adsorption sites of the matrix. For the extraction of 2-cyclopentyl-cyclopentanone, water was selected as a displacer primarily because it is known to be easily adsorbed by polyamide 6.6 and it is thus expected to be an efficient displacer [67]. The effect of water on the migration of 2-cyclopentyl-cyclopentanone was studied by performing extractions with and without water. Figure 5 shows the extraction profiles of Zytel polyamide extracted for 30 min at 80 °C without water and for 20 min at 90 °C with water displacer.

Polyamide 6.6 samples were incubated between 0 and 25 h at 80 °C or between 0 and 10 h at 90 °C before the HS-SPME [67]. A total of 250 μl of water was added to the samples extracted at 90 °C. The curves shown in Fig. 5 point out the significant reduction in time required for optimal recovery when incubating at higher temperature and in the presence of a displacer. At 80 °C, 10 h was required to reach the optimum recovery. This is again due to the hydrogen bonding between the analyte and the polar polyamide 6.6 matrix, which causes slow migration of the analyte. When water was added as a displacer, and the extraction was performed at 90 °C, the optimum recovery was reached already after 1 h. The significant reduction in time required demonstrates the ability of water to break the hydrogen bonding between the analyte and the matrix. After addition of water as a displacer, a MHS-SPME extraction plot with good linearity and an average correlation coefficient of 0.991 was obtained. The plot was linear throughout the multiple extractions and no tendency to adsorption was noticed. Using the given conditions for MHS-SPME, the concentration of 2-cyclopentyl-cyclopentanone in Zytel was measured to be 15.61 μg/g.

To validate the developed MHS-SPME method, the concentration of 2-cyclopentyl-cyclopentanone in five different polyamide 6.6 samples was measured by MHS-SPME and the results were compared to the concentrations determined by the previously developed MAE method [67]. In general, the amounts determined by MHS-SPME and MAE agreed well. However, the concentrations determined by MHS-SPME were up to 30% higher than the concentrations measured after MAE. As it was shown that the developed MHS-SPME method eliminated the matrix effects, it is likely that the con-

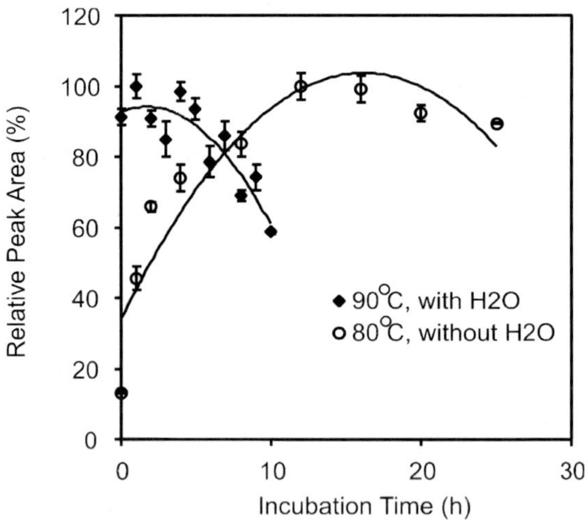

Fig. 5 The extraction profiles of 25 mg Zytel extracted for 30 min at 80 °C and 30 mg Zytel extracted for 20 min at 90 °C together with 250 μl water as displacer. The samples were incubated for up to 25 h prior to extraction. Reprinted from [67] with permission of Elsevier. © Elsevier (2004)

centrations measured by MHS-SPME are the true concentrations and the recovery of the MAE was around 70–85%.

4.6 Rapid Assessment of Volatile Content Under Non-Equilibrium Conditions

The content of volatiles in polymers is routinely assessed by the polymer producing and processing industry. Generally, HS-GC is used under non-equilibrium conditions and absolute contents are not measured. The performance of HS-SPME, compared to traditional HS, for assessing the 2-cyclopentyl-cyclopentanone content in polyamide 6.6 under non-equilibrium conditions was evaluated by extracting the analyte from five polyamide 6.6 samples using traditional HS and HS-SPME at 80 °C and 120 °C [66]. The influences of different parameters on the extractions were studied. Table 2 shows the amount of 2-cyclopentyl-cyclopentanone extracted from five polyamide 6.6 samples using traditional HS extraction and HS-SPME at 80 °C and 120 °C.

Under non-equilibrium conditions, sample size, extraction temperature and extraction method (HS or HS-SPME) strongly affected the amount of 2-cyclopentyl-cyclopentanone emitted from polyamide 6.6 [66]. The amount

Table 2 The amount of 2-cyclopentyl-cyclopentanone in the five different polyamide 6.6 samples measured by non-equilibrium HS-SPME and HS-extraction at 80 °C and 120 °C showing the large effect of extraction parameters on the correctness of the results. Reprinted from [66] with permission of Elsevier. © Elsevier (2004)

$n = 4$	HS						HS-SPME			
	Granules		*Powder*				*Powder*			
	120 °C, 5 h $r^2 = 1.000$		120 °C, 45 min $r^2 = 0.995$		80 °C, 45 min $r^2 = 0.993$		120 °C, 45 min $r^2 = 0.995$		80 °C, 45 min $r^2 = 0.986$	
Polyamide	Mean (ng/g)	STD (RSD)	Mean (ng/g)	STD (RSD)	Mean (ng/g)	STD (RSD)	Mean (ng/g)	STD (RSD)	Mean (ng/g)	STD (RSD)
Zytel	343.6	13.0 (3.8)	911.2	22.6 (2.5)	136.9	4.5 (3.3)	5144.4	255.4 (5.0)	606.6	86.7 (14.3)
Aldrich	293.2	11.3 (3.8)	922.1	33.2 (3.6)	180.8	25.3 (14.0)	2998.2	457.8 (15.3)	570.0	80.0 (14.0)
Base	6.2	1.4 (23.2)	55.9	17.5 (31.3)	20.8	8.8 (42.2)	250.7	25.9 (10.3)	44.7	0.6 (1.4)
Recovered	154.0	7.4 (4.8)	522.7	23.3 (4.5)	65.4	8.3 (12.7)	923.4	42.7 (4.6)	219.2	32.0 (14.6)
Compound	56.0	2.3 (4.1)	161.3	28.0 (17.3)	15.6	0.5 (3.3)	300.7	33.7 (11.2)	92.3	8.2 (8.9)

measured after HS-SPME for all the samples was two to five times higher than the amount measured after HS. Both HS and HS-SPME showed that the amounts emitted at 120 °C are approximately three to ten times higher than the amounts emitted at 80 °C. The large effect of sample size on the migration rate and equilibrium time was also demonstrated as the amount of 2-cyclopentyl-cyclopentanone that had migrated from the polyamide granules after 5 h of incubation at 120 °C was lower than the amount that had migrated from the polyamide powder after only 45 min of incubation at 120 °C.

2-cyclopentyl-cyclopentanone was the single most abundant compound in the chromatograms after HS and SPME extractions from the Aldrich, Zytel and Base polyamide materials representing more than 95% of the total peak area. In the extractions from the recovered material, a large amount of hydrocarbons were found and 2-cyclopentyl-cyclopentanone peak area represented less than 10% of the total peak area. This indicates that the recovered polyamide 6.6 either contains lubricants or is contaminated by polyolefins. The compound material consisted of 47% recovered material, 20% base material, 30% glass-fibers and 3% additives. As the base material contained very low amounts of 2-cyclopentyl-cyclopentanone, it was expected that the amount of 2-cyclopentyl-cyclopentanone emitted from the compound material would be approximately half of the amount emitted from the recovered material. However, the measured amounts were only 25–40% of the amount emitted from the recovered material. This is explained by degradation of 2-cyclopentyl-cyclopentanone during compounding and was also confirmed in the study concerning in-plant recycling of polyamide 6.6. The amount 2-cyclopentyl-cyclopentanone emitted from the samples measured by the non-equilibrium headspace methods developed was found to be in good correlation with the total amount in the samples measured by MAE. This shows that it is possible to estimate the amount of 2-cyclopentyl-cyclopentanone in polyamide 6.6 using headspace measurements under non-equilibrium conditions and without compensating for volume- or matrix effects. The headspace techniques can thus be applied to rapidly compare or quality control different samples with respect to their volatile contents.

5 SPME in Quality Control of Recycled Polymers

The deteriorating of plastic products with time is well known and calls for careful studies of polymer durability and long-term properties. In addition, with increasing demands on recycling of plastics, the effects of repeated processing on polymer durability is an increasingly important issue. The durability of polyamide 6.6 is closely linked to its thermal history. It has been shown that repeated injection moulding and the addition of glass-fibers reduces the oxidative stability of polyamide 6.6 [72]. In addition, the short-term mechanical performance of glass-fiber-reinforced polyamide 6.6 was found to decrease by approximately 10% due to grinding and reprocessing [73]. Polyamide 6.6 could, however, be reprocessed without significant deterioration of the short-term mechanical properties if the regrind is compounded with at least 70% of virgin material [74]. Although the short-term properties are only little affected by repeated processing, larger effects may be observed with respect to long-term properties. During the thermo-oxidative ageing of non-reinforced and glass-fiber reinforced recycled polyamide 6.6, faster increase in carbonyl index and simultaneous decreases in melting peak temperature and elongation at break were observed for reprocessed rather than virgin samples [75, 76]. When a regrind level of only 25% was used in the samples, no significant influence on the long-term properties could be detected and the deterioration rate was similar to that of the virgin samples [77]. It was hence concluded that recycled polyamide 6.6 is more susceptible to degradation during service life than virgin polyamide 6.6.

5.1 Thermo-Oxidation of In-plant Recycled Polyamide 6.6

HS-SPME is a usable technique also for polymer durability assessment as volatile content and degradation product patterns change during reprocessing and long-term aging. Repeated processing was shown to increase the formation of low-molecular-weight compounds during subsequent long-term thermo-oxidation of polyamide 6.6 [65]. Figures 6 and 7 show the chromatograms of compounds extracted from virgin and recycled polyamide 6.6 after 25 and 1200 h of thermo-oxidation at 100 °C, respectively.

After 25 h of ageing at 100 °C, nine compounds were extracted from virgin material whereas 12 compounds were extracted from the material recycled three times. After 1200 h of oxidation the number of compounds extracted had increased to 14 in the case of the virgin material compared to 16 in the case of the material recycled three times. It should be observed that the peak areas in the chromatograms depend on the distribution of analytes in the fiber coating and do not reflect the absolute amount of analytes in the polyamide 6.6. The distribution of analytes in the fiber is determined by the individual analytes partitioning coefficients between the condensed analyte, its headspace and the fiber coating. As an example, Fig. 8 shows the peak areas after triplicate extractions of some selected analytes using different fiber coatings. PDMS/DVB fiber was somewhat more effective for extraction of 2-cyclopentyl-cyclopentanone, ethyl-succinimide and collidine (2,4,6-trimethyl-pyridine) than for extraction of the other analytes. Although equal amounts of all analytes were used for all the extractions, the peaks corresponding to caprolactame, glutarimide, pentanamide and 3-picoline were significantly smaller compared to the other analytes. This may not necessarily be due only to different partitioning coefficients, as also differences in vapor

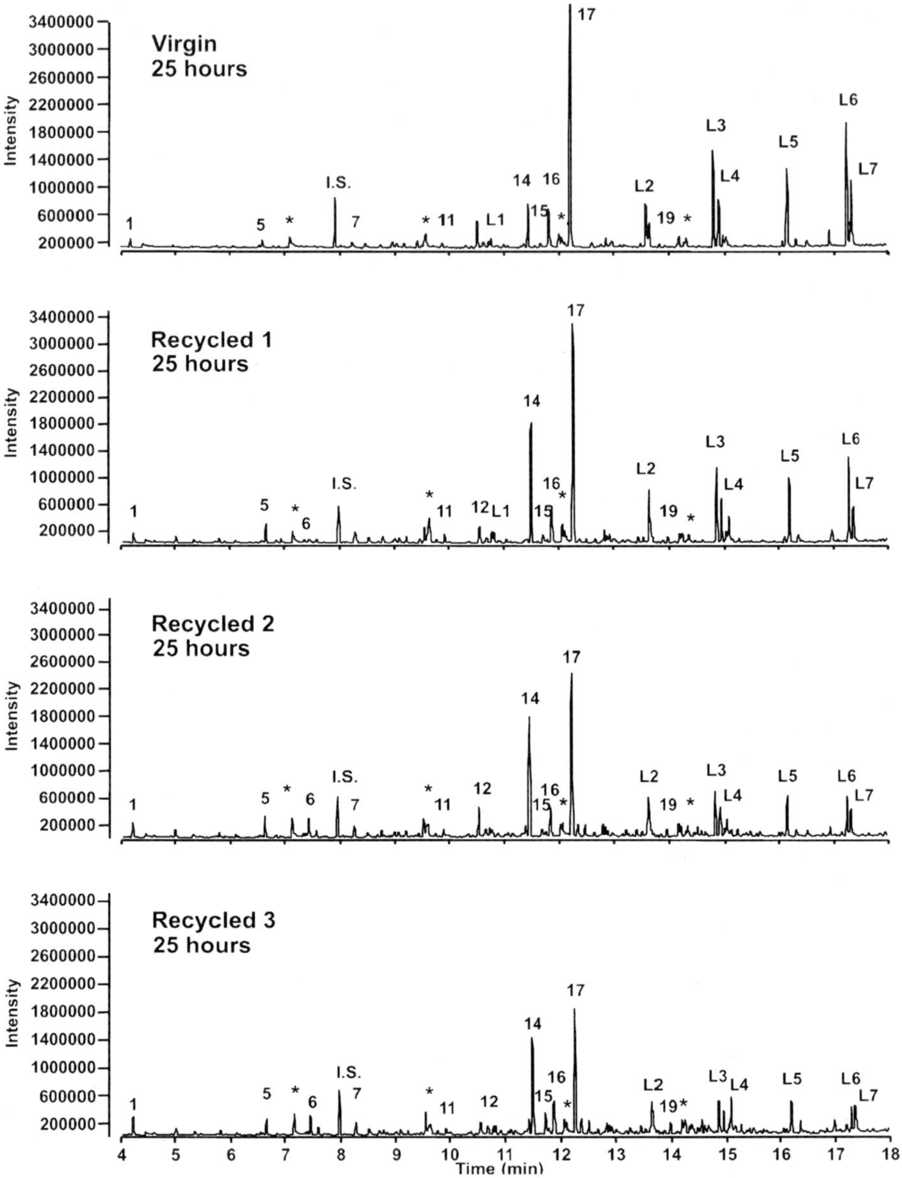

pressures are expected to influence the extracted amounts. In agreement with Fig. 1, higher recoveries were achieved with the CAR/DVB fiber than with the PDMS/DVB fiber, but the repeatability was significantly lower.

The identified compounds were classified into four groups: cyclic imides, pyridines, chain fragments and cyclopentanones, as presented in Fig. 9 [65]. Of these compounds, the cyclic imides, pyridines and chain fragments gen-

◀ **Fig. 6** Chromatograms of extractions from virgin and recycled polyamide 6.6 after 25 h of thermo-oxidation at 100 °C. Identity of the peaks is 1 = cyclopentanone, 2 = 2-methylpyridine, 3 = pentanoic acid, 4 = butanamide, 5 = 2-ethylcyclopentanone, 6 = 2,4,6-trimethylpyridine, 7 = pentanamide, 8 = 3-(1-methylethyl)pyridine, 9 = 2-butylpyridine, 10 = N,N-hexamethylenebisformamide, 11 = 2-butylcyclopentanone, 12 = glutarimide, 13 = 1-propyl-2,5-pyrrolidinedione, 14 = 2-pentylcyclopentanone, 15 = caprolactam, 16 = azepane-2,7-dione, 17 = 2-cyclopentyl-cyclopentanone, 18 = 1-butyl-2,5-pyrrolidinedione, 19 = 1-pentyl-2,5-pyrrolidinedione, 20 = 2-butyl-3,5-dimethylethylpyridine, L1–L7 = linear C13-C17 alkanes and alkenes from lubricant, * = silicone from septa used to seal vials. Reprinted from [65] with permission of John Wiley & Sons, Inc. © John Wiley & Sons, Inc (2002)

erally increased during the thermo-oxidation and reprocessing, while the amount of cyclopentanones decreased. Five cyclic imides were identified: 2,6-piperidinedione (glutarimide), 1-propyl-2,5-pyrrolidinedione, azepane-2,7-dione, 1-butyl-2,5-pyrrolidinedione and 1-pentyl-2,5-pyrrolidinedione. As shown in Fig. 10, the total amount of cyclic imides extracted clearly increased as a function of thermo-oxidation time and the formation rate increased as a function of number of reprocessing times. 1-pentyl-2,5-pyrrolidinedione was the compound that showed the most prominent increase in abundance due to thermo-oxidation and after 1200 h its relative abundance had increased from 4 to 106 in virgin polyamide 6.6 and from 30 to 453 in polyamide 6.6 recycled three times. At the end of the oxidation period, 1-pentyl-2,5-pyrrolidinedione was the most abundant compound in both virgin and recycled materials. However, significantly larger amounts of 1-pentyl-2,5-pyrrolidinedione were extracted from recycled polyamide 6.6 compared to virgin polyamide 6.6 and the increase in abundance corresponded almost linearly to the number of recycling steps the material had been subjected to.

The identified pyridines were: 2-methyl-pyridine, 2,4,6-trimethyl-pyridine, 3-(1-methylethyl)-pyridine, 2-butyl-pyridine and 2-butyl-3,5-dimethylethyl-pyridine [65]. Considerably larger amounts of pyridines were extracted from recycled than virgin polyamide 6.6. The amount of pyridines increased during the aging in all the materials, but the increase was faster for reprocessed materials. The amount of chain fragments also increased due to thermo-oxidation and repeated processing. Recycling facilitated the formation of chain fragments only after relatively long oxidation times, as a distinct difference between virgin and recycled materials was observed first after 500 h of ageing. The effect of repeated recycling on formation of chain fragments was quite small however.

Cyclopentanone and four cyclopentanone derivatives were identified: 2-ethyl-cyclopentanone, 2-butyl-cyclopentanone, 2-pentyl-cyclopentanone and 2-cyclopentyl-cyclopentanone [65]. At the beginning of oxidation, i.e. after 25 h of ageing, 2-cyclopentyl-cyclopentanone was the most abundant compound in the extractions from both virgin and recycled material. In contrast to the other compounds extracted, the amount of cyclopentanones and espe-

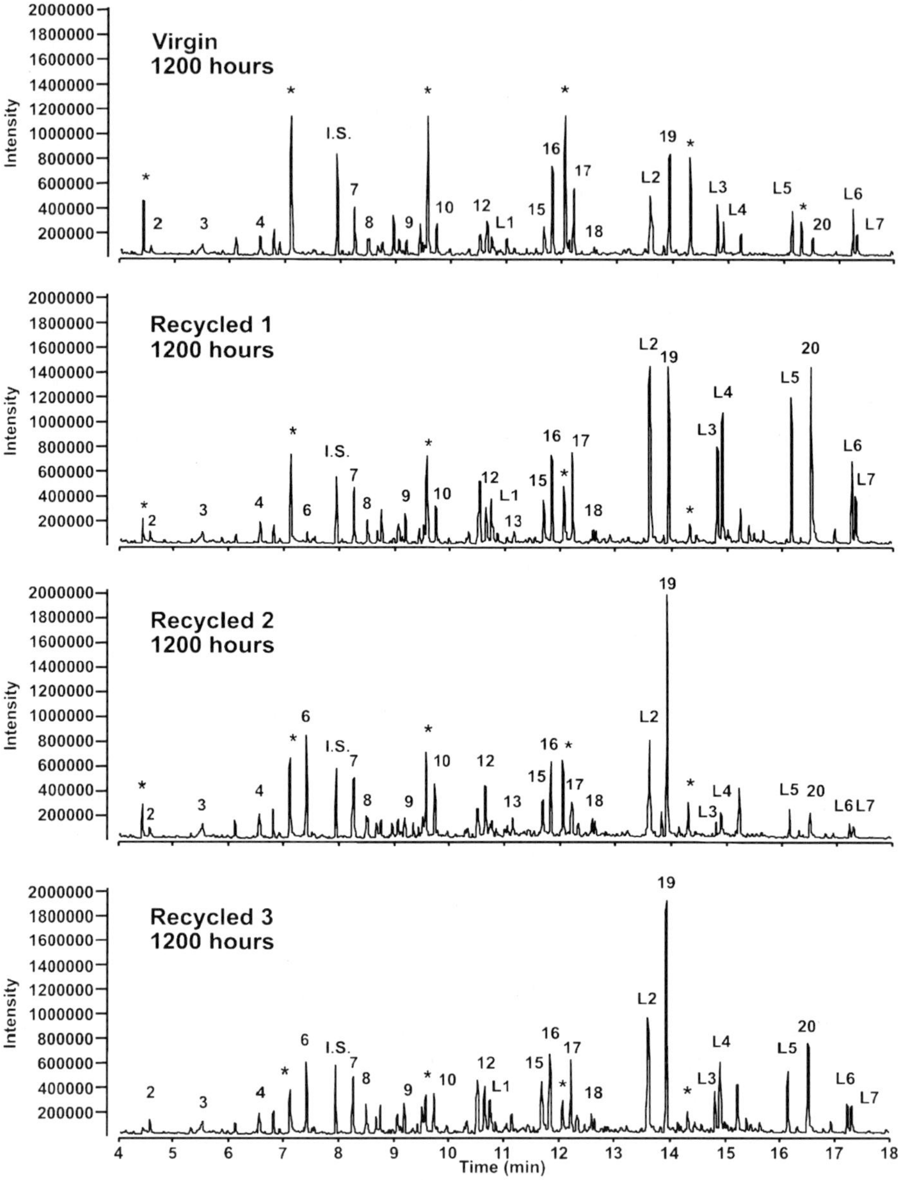

cially 2-cyclopentyl-cyclopentanone decreased during the thermo-oxidation and reprocessing, as shown in Fig. 11. After 1200 h, only 6% of the original amount of 2-cyclopentyl-cyclopentanone could be extracted from the virgin material. Furthermore, the largest amount of 2-cyclopentyl-cyclopentanone was extracted from virgin material; approximately 3.5 times more than from three times recycled material. Repeated processing reduced the amount of

◀ **Fig. 7** Chromatograms of extractions from virgin and recycled polyamide 6.6 after 1200 h of thermo-oxidation at 100 °C. Identity of the peaks is 1 = cyclopentanone, 2 = 2-methylpyridine, 3 = pentanoic acid, 4 = butanamide, 5 = 2-ethylcyclopentanone, 6 = 2,4,6-trimethylpyridine, 7 = pentanamide, 8 = 3-(1-methylethyl)pyridine, 9 = 2-butylpyridine, 10 = N,N-hexamethylenebisformamide, 11 = 2-butylcyclopentanone, 12 = glutarimide, 13 = 1-propyl-2,5-pyrrolidinedione, 14 = 2-pentylcyclopentanone, 15 = caprolactam, 16 = azepane-2,7-dione, 17 = 2-cyclopentyl-cyclopentanone, 18 = 1-butyl-2,5-pyrrolidinedione, 19 = 1-pentyl-2,5-pyrrolidinedione, 20 = 2-butyl-3,5-dimethylethylpyridine, L1–L7 = linear C13-C17 alkanes and alkenes from lubricant, * = silicone from septa used to seal vials. Reprinted from [65] with permission of John Wiley & Sons, Inc. © John Wiley & Sons, Inc (2002)

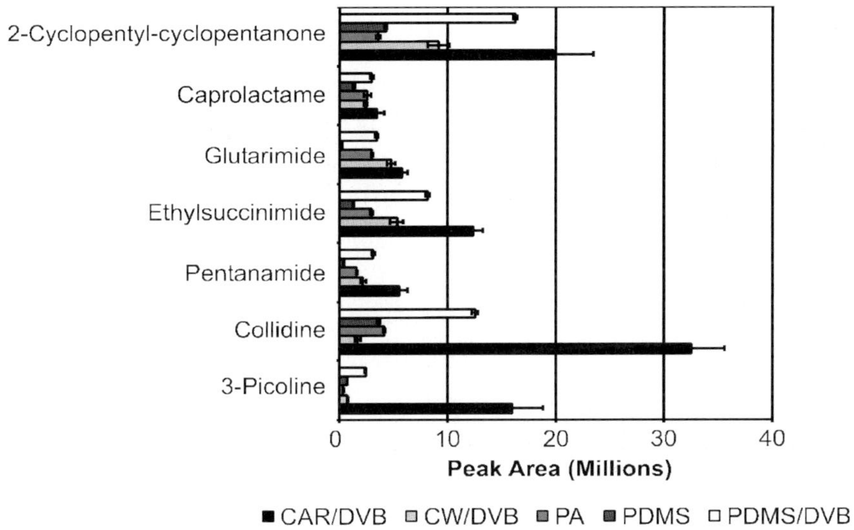

Fig. 8 100 ng of some selected compounds extracted using different fiber coatings

2-cyclopentyl-cyclopentanone, however, the amount of other cyclopentanones increased. Hence, it was concluded that 2-cyclopentyl-cyclopentanone was formed during the polymerization of polyamide 6.6 and it degraded into lower-molecular-weight cyclopentanones during processing.

5.2 Long-Term Properties of In-plant Recycled Polyamide

Identification of the formed degradation products serves as a means to establish the degrading reactions that ultimately lead to material failure. In addition to following the formation of low-molecular-weight compounds, the increased susceptibility to thermo-oxidation due to repeated recycling was also studied with more traditional techniques including tensile testing, differential scanning calorimetry (DSC) and Fourier transform infrared spec-

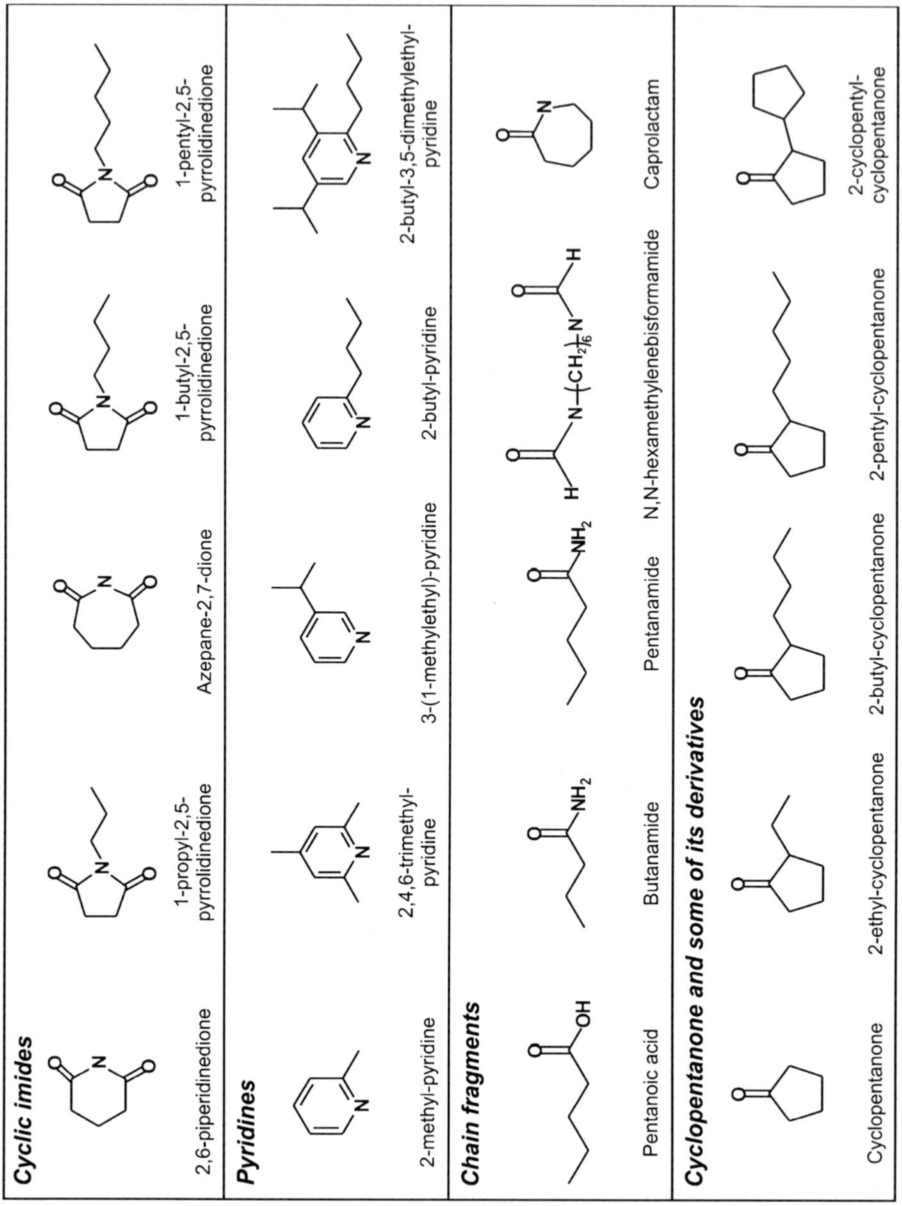

Fig. 9 Classification of identified compounds in virgin and recycled polyamide 6.6 where compounds in each group share a structural feature. Reprinted from [65] with permission of John Wiley & Sons, Inc. © John Wiley & Sons, Inc (2002)

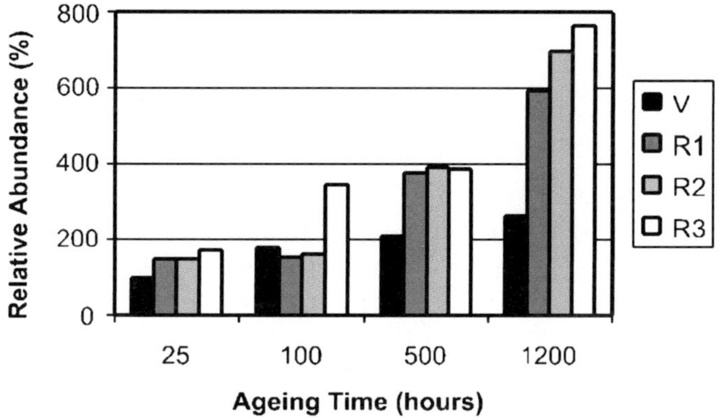

Fig. 10 Changes in extracted amount of succinimides from virgin and recycled polyamide 6.6 during ageing at 100 °C for 1200 h. Reprinted from [65] with permission of John Wiley & Sons, Inc. © John Wiley & Sons, Inc (2002)

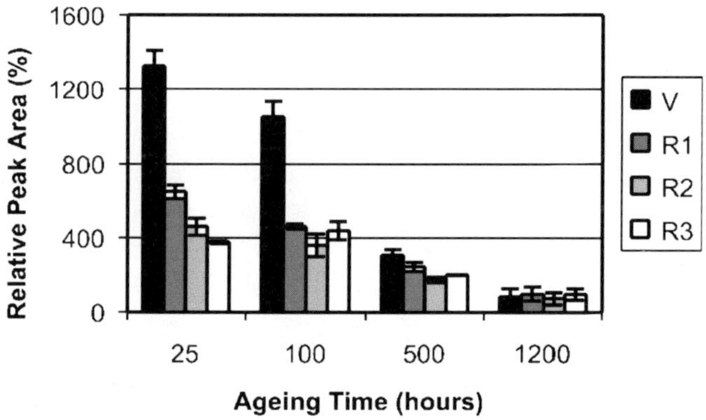

Fig. 11 Decrease in extractable amount of 2-cyclopentyl-cyclopentanone from virgin and recycled polyamide 6.6 during thermo-oxidation at 100 °C. Reprinted from [65] with permission of John Wiley & Sons, Inc. © John Wiley & Sons, Inc (2002)

troscopy (FTIR) [65]. A distinct difference was evident between virgin and recycled materials also from a visual examination of the samples during ageing as the latter were yellowed earlier and more severely.

During the first 100 h of oxidation there were no significant differences between the samples, i.e. they all seemed unaffected by thermo-oxidation and no influence of repeated processing was found [65]. However, after 500 h of thermo-oxidation, all recycled samples showed a marked deterioration in tensile strength whereas the virgin material remained unaffected. The magnitude of the decrease in tensile strength corresponded well to the number of times the material was extruded. After 1200 h of thermo-oxidation the deterioration in mechanical properties was even more pronounced and at this point the polyamide 6.6 recycled three times retained only 20% of its original tensile strength compared to 75% for the virgin material. The corresponding percentages for once- and twice-recycled material were 50 and 40%, respectively. DSC showed that the melt peak temperature of polyamide 6.6 decreased due to thermo-oxidation. Significant differences between the different samples were observed first after 500 h of thermo-oxidation when the melting peak temperature of virgin and once recycled material was found to be somewhat higher than the melting peak temperature for materials recycled two and three times. No significant influence of repeated recycling could be seen on the amount of carbonyl groups at the surface of the materials. The large numbers of carbonyl groups already present in the polyamide backbone makes it difficult to measure the changes in carbonyl groups due to oxidation. These results demonstrated the value of the SPME technique as a more sensitive tool compared to DSC, FTIR and tensile testing for durability assessment and early degradation detection.

5.2.1 Correlation Between Formation of Degradation Products and Changes in Tensile Strength

A good correlation between the formation of degradation products and the simultaneous reduction in tensile strength during thermo-oxidation was observed [65]. Figures 12 and 13 show the increase in the relative amount of the most abundant thermo-oxidation product, 1-pentyl-2,5-pyrrolidinedione and the simultaneous decrease in tensile strength for virgin and once recycled material, respectively.

From Figs. 12 and 13 it is clear that the largest reduction in tensile strength coincided with the large increase in the amount of 1-pentyl-pyrrolidinedione.

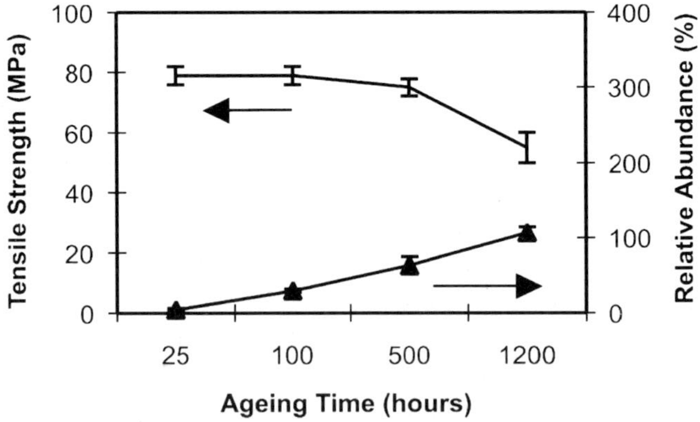

Fig. 12 Correlation between formation of 1-pentyl-2,5-pyrrolidinedione and simultaneous decrease in tensile strength for virgin polyamide 6.6 during thermo-oxidation at 100 °C. Reprinted from [65] with permission of John Wiley & Sons, Inc. © John Wiley & Sons, Inc (2002)

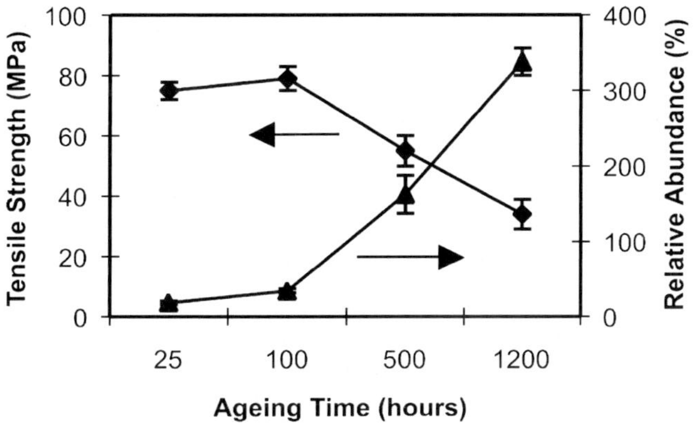

Fig. 13 Correlation between formation of 1-pentyl-2,5-pyrrolidinedione and simultaneous decrease in tensile strength for once-recycled polyamide 6.6 during thermo-oxidation at 100 °C. Reprinted from [65] with permission of John Wiley & Sons, Inc. © John Wiley & Sons, Inc (2002)

This was particularly prominent for the once-recycled material. Furthermore, the previously discussed increased susceptibility to thermo-oxidation due to repeated processing was also clearly demonstrated. For virgin material, the changes in both tensile strength and abundance of 1-pentyl-2,5-pyrrolidinedione were relatively modest even after 1200 h of thermo-oxidation. However for the once-recycled material, the changes were already quite large after 500 h of thermo-oxidation. Also, the number and amount of the other thermo-oxidation products increased simultaneously, e.g. in the recycled samples a large amount of chain fragments were formed between 100 and 500 h of oxidation, which coincided with the large decrease in mechanical properties.

5.3 Screening for Brominated Flame Retardants in Polyamide Collected for Recycling

For quality control reasons, rapid screening methods are needed to identify the volatiles in polymeric materials collected for recycling. HS-SPME-GC-MS was shown to be a fast and sensitive method to screen for brominated flame retardants in recycled polyamide materials [78]. HS-SPME effectively extracted several brominated compounds, all possible degradation products from the common flame-retardant Tetrabromobisphenol A from recycled polyamide 6.6. Furthermore, the high extraction capacity of the PDMS/DVB stationary phase towards aromatic compounds was demonstrated, as the HS-SPME-GC-MS method allowed the extraction and identification of brominated benzenes, from a complex matrix only containing trace amounts of analytes. In addition, degradation products from an antioxidant, a hindered phenol, were extracted. Figure 14 shows a typical chro-

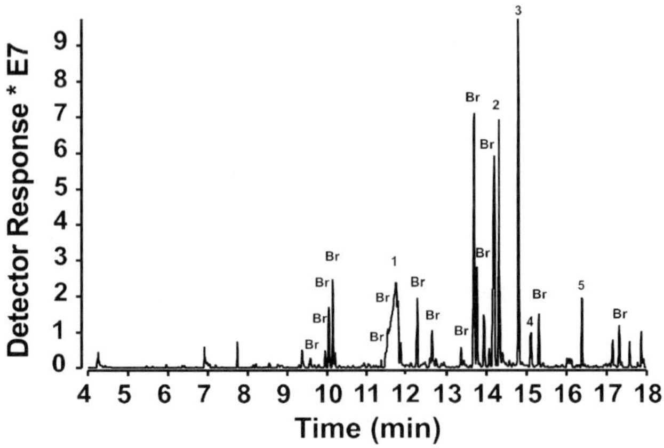

Fig. 14 GC-MS chromatogram showing the compounds extracted by SPME from polyamide 6.6 collected for recycling. The identities of the numbered peaks are: Br = brominated benzene; 1 = caprolactam; 2–4 = hindered phenols; 5 = diphenylamine. Reprinted from [78] with permission of John Wiley & Sons, Inc. © John Wiley & Sons, Inc (2003)

matogram of the volatile compounds migrating from polyamide 6.6 collected for recycling.

5.4
Quality Control of Polypropylene and Polyamide 6 Containing Recycled Phenol-formaldehyde Resin as a Functional Filler

Large amounts of phenolic prepreg scrap are generated in the production of grinding wheels when circular discs, used as reinforcement layers in the wheels, are punched out from continuous impregnated glass-fiber mats. There are strong incentives to recycle the waste: approximately 20–30% of the woven laminates entering the cutting process are discarded as waste [79]. In addition, the presence of phenol makes the uncured resin (due to the risk of monomer leakage) unsuitable for disposal in landfills. We have shown that in-plant scrap of woven glass fibers impregnated with phenol-formaldehyde resin can be re-used as functional filler with reinforcing and antioxidant effects in polypropylene (PP) and polyamide 6 (PA 6) [40, 80, 81] headspace-gas chromatography-mass spectrometry (HS-GC-MS) showed that PP/glassfiber phenol-formaldehyde composites emitted somewhat larger amounts of volatile compounds compared to the reference PP/glass fiber composites, while the amount of volatile components emitted from PA6/glassfiber phenol-formaldehyde composites was similar to the reference PA6/glass fiber composites [40]. The comparison of chromatographic emission-profiles from PP and PA6 reference samples to prepreg-filled PP and PA6 composites shows that the only significant differences in emissions are the presence of phenol and an unidentified cyclic compound in the prepreg composites, which are not emitted from the reference materials. The amount of phenol emitted from PA6/prepreg was smaller that the amount emitted from PP/prepreg. This is probably due to the slower diffusion rate of phenol in PA6 caused by stronger secondary forces, e.g. hydrogen bonding between phenol and PA6 and could also be influenced by the higher processing temperature used for PA6 composites.

6 Conclusions

HS-SPME is a very useful tool in polymer analysis and can be applied for absolute and semi-quantitative determination of the volatile content in polymers, for degradation studies, in the assessment of polymer durability, for screening tests and for quality control of recycled materials. For quantitative determination of volatiles in polymers, SPME can be combined with multiple headspace extraction to remove the matrix effects. If the linearity of the MHS-SPME plot has been verified, the number of extractions can be reduced to two, which considerably reduces the total analysis time. Advantages of MHS-SPME compared to MAE are its higher sensitivity, the small sample amount required, solvent free nature and if an autosampler is used a low demand of labor time. In addition, if the matrix effects are absent, the recovery will always be 100%. This is valuable compared to other techniques for extracting volatiles in polymers in which the recovery should be calculated from the extraction of spiked samples, which are very difficult to produce in the case of polymeric materials.

Strong interactions between the polar matrix and polar analytes may lead to extremely long equilibrium times and errors in quantitation even when the MHS technique is used. In these cases, a displacer may be added to break the interactions between the matrix and analyte. Polar 2-cyclopentyl-cyclopentanone could be quantitatively determined in polar polyamide 6.6 by MHS-SPME if water was added as a displacer to break the hydrogen bonding between 2-cyclopentyl-cyclopentanone and polyamide. The addition of water also significantly reduced the equilibrium time. A correlation was found between the amount of 2-cyclopentyl-cyclopentanone emitted from polyamide 6.6 and the total amount of 2-cyclopentyl-cyclopentanone in the material. This correlation enables rapid assessment of the 2-cyclopentyl-cyclopentanone content using headspace techniques under non-equilibrium conditions. The analysis time is significantly reduced if the polymer samples are milled to a powder prior to extraction.

HS-SPME-GC-MS and HS-GC-MS were shown to be good tools to quality control recycled materials. The short-term properties of un-stabilized polyamide 6.6 were only slightly affected by in-plant recycling. However, during accelerated ageing in air at 100 °C, larger amounts of low-molecular-degradation compounds and more severe deterioration of tensile strength was observed for recycled material. Furthermore, degradation-induced changes in material properties occurred earlier for recycled material than for virgin material, thus leading to faster deteriorating of long-term material properties. Of the degradation products, N-alkyl-succinimides showed the largest increase in abundance due to ageing. The increase in the amount of N-alkyl-succinimide coincided with the decrease in tensile strength. N-alkyl-succinimide could thus be used as an indicator product to determine if the material still has useful properties. HS-SPME-GC-MS was a sensitive tool for monitoring the thermo-oxidative degradation of polyamide 6.6 as it detected degradation induced-changes in the product pattern during the early stages of ageing when DSC, tensile testing and FTIR could not yet detect any changes. HS-SPME was also a very good technique for the screening of brominated flame-retardant in polyamide materials collected for recycling.

References

1. Luque de Castro MD, Garcia-Ayuso LE (1998) Anal Chim Acta 369:1
2. Eskilsson CS, Björklund E (2000) J Chromatogr A 902:227
3. Lou XW, Janssen HG, Cramers CA (1997) Anal Chem 69:1598
4. Vandenburg HJ, Clifford AA, Bartle KD, Carroll J, Newton I, Garden LM, Dean JR, Costley CT (1997) Analyst 122:R101
5. Vandenburg HJ, Clifford AA, Bartle KD, Carroll J, Newton I (1999) Analyst 124:1707
6. Schmidt S, Blomberg L, Waennman T (1989) Chromatographia 28:400
7. Bart JCJ (2003) Polym Degrad Stab 82:197
8. Hachenberg H, Schmidt AF (1977) Gas Chromatographic Headspace Analysis. Wiley, New York
9. Kolb B, Ettre LS (eds) (1997) Static Headspace-Gas Chromatography: Theory and Practice. Wiley VCH, New York
10. Ettre LS (2001) Adv Exp Med Biol 488:9
11. Hagman A, Jacobsson S (1987) J Chromatogr 395:271
12. Hagman A, Jacobsson S (1988) J High Res Chromatogr 11:830
13. Hagman A, Jacobsson S (1989) Anal Chem 61:1202
14. Hodgson SC, O'Connor MJ, Casey RJ, Bigger SW (1998) J Agric Food Chem 46:1397
15. Bart JCJ (2001) Polym Test 20:729
16. Bovijn L, Pirotte J, Berger A (1958) In: Desty DH (ed) Gas Chromatography. Proc 2nd Symposium, Amsterdam p 310
17. Romano SJ, Renner JA, Leitner PM (1973) Anal Chem 45:2327
18. Shanks RA (1975) Scan 6:20
19. ASTM, D 4526 - Determination of Volatiles in Polymers by Static Headspace Gas Chromatography (2001)
20. Arthur CL, Pawliszyn J (1990) Anal Chem 62:2145
21. Zhang Z, Pawliszyn J (1993) Anal Chem 65:1843
22. Gorecki T, Yu X, Pawliszyn J (1999) Analyst 124:643
23. Lord H, Pawliszyn J (2000) J Chromatogr A 885:153
24. Gorecki T, Pawliszyn J (1997) Analyst 122:1079

25. Gorecki T, Khaled A, Pawliszyn J (1998) Analyst 123:2819
26. Ai J (1997) Anal Chem 69:3260
27. Ai J (1997) Anal Chem 69:1230
28. Ai J (1998) Anal Chem 70:4822
29. Hakkarainen M, Albertsson A-C, Karlsson S (1997) J Environ Polym Degrad 5:67
30. Ezquerro Ò, Pons B, Tena MT (2002) J Chromatogr A 963:381
31. Fabre M, Aubry V, Guichard E (2002) J Agric Food Chem 50:1497
32. Fäldt J, Eriksson M, Valterova I, Borg-Karlson A-K (2000) Z Naturforsch C 55:180
33. ASTM, D 6889 - Fast Screening for Volatile Organic Compounds in Water Using Solid-Phase Microextraction (SPME) (2003)
34. ASTM, D 6520 - Solid-Phase Microextraction (SPME) of Water and its Headspace for the Analysis of Volatile and Semi-Volatile Organic Compounds (2000)
35. ASTM, D 3749 - Residual Vinyl Chloride Monomer in Poly(Vinyl Chloride) Resins by Gas Chromatography Headspace Technique (2002)
36. Sala C, Mestres M, Marti MP, Busto O, Guasch J (2000) J Chromatogr A 880:93
37. Sala C, Mestres M, Marti MP, Busto O, Guasch J (2002) J Chromatogr A 953:1
38. Scarlata CJ, Ebeler SE (1999) J Agric Food Chem 47:2505
39. Meininghaus R, Fuhrmann F, Salthammer T (1996) Fresen J Anal Chem 356:344
40. Gröning M, Eriksson H, Hakkarainen M, Albertsson A-C (2006) Polym Degrad Stab 91:1815
41. Chai XS, Hou QX, Schork FJ (2004) J Chromatogr A 1040:163
42. ASTM, F 2013 - Determination of Residual Acetaldehyde in Polyethylene Terephtalate Using an Automated Static Head-Space Sampling Device and a Capillary GC with a Flame Ionization Detector (2001)
43. ASTM, D 4443 - Determining Residual Vinyl Chloride Monomer Content in PPB Range in Vinyl Chloride Homo- and Co-Polymers by Headspace Gas Chromatography (2002)
44. ASTM, D 4322 - Residual Acrylonitrile Monomer Styrene-Acrylonitrile Copolymers and Nitrile Rubber by Headspace Gas Chromatography (2001)
45. Eriksson M, Fäldt J, Dalhammar G, Borg-Karlson A-K (2001) Chemosphere 44:1641
46. Clark TJ, Bunch JE (1997) J Agric Food Chem 45:844
47. Yang L, Colombini V, Maxwell P, Mester Z, Sturgeon RE (2003) J Chromatogr A 1011:135
48. Kolb B, Pospisil P (1977) Chromatographia 10:705
49. Rohrschneider L (1971) Fresen Z Anal Chem 255:345
50. Kolb B, Pospisil P, Auer M (1981) J Chromatogr 204:371
51. Kolb B (1982) Chromatographia 15:587
52. Venema A (1988) J High Res Chromatogr 11:128
53. Venema A (1986) J High Res Chromatogr 9:637
54. Kolb B, Ettre LS (1991) Chromatographia 32:505
55. Hakkarainen M (2007) J Biochem Biophys Methods 70:229
56. Tena MT, Carrillo JD (2007) TRAC-Trend Anal Chem 26:206
57. Hansson E, Hakkarainen M (2006) J Chromatogr A 1102:91
58. Kolb B, Pospisil P, Auer M (1984) Chromatographia 19:113
59. Milana MR, Maggio A, Denaro M, Feliciani R, Gramiccioni L (1991) J Chromatogr 552:205
60. Ezquerro Ó, Ortiz G, Pons B, Tena MT (2004) J Chromatogr A 1035:17
61. Wenzl T, Lankmayr EP (2002) Anal Bioanal Chem 372:649
62. Wenzl T, Lankmayr EP (2001) J Sep Sci 24:885
63. Wenzl T, Lankmayr EP (2000) J Chromatogr A 897:269

64. Gröning M, Hakkarainen M (2001) J Chromatogr A 932:1
65. Gröning M, Hakkarainen M (2002) J Appl Polym Sci 86:3396
66. Gröning M, Hakkarainen M (2004) J Chromatogr A 1052:151
67. Gröning M, Hakkarainen M (2004) J Chromatogr A 1052:61
68. Soto-Valdez H, Gramshaw JW, Vandenburg HJ (1997) Food Addit Contam 14:309
69. Soto-Valdez H, Gramshaw JW (2000) J Mater Sci Lett 19:823
70. Gramshaw JW, Soto-Valdez H (1998) Food Addit Contam 15:329
71. Allen NS, Harrison MJ (1985) Eur Polym J 21:517
72. Eriksson PA, Albertsson A-C, Eriksson K, Månson JAE (1998) J Therm Anal Calorim 53:19
73. Eriksson PA, Albertsson A-C, Boydell P, Eriksson K, Månson JAE (1996) Polym Composit 17:823
74. Eriksson PA, Albertsson A-C, Boydell P, Prautzsch G, Månson JAE (1996) Polym Composit 17:830
75. Eriksson PA, Boydell P, Månson JAE, Albertsson A-C (1997) J Appl Polym Sci 65:1631
76. Eriksson PA, Boydell P, Eriksson K, Månson JAE, Albertsson A-C (1997) J Appl Polym Sci 65:1619
77. Eriksson PA, Albertsson A-C, Boydell P, Månson JAE (1998) Polym Eng Sci 38:348
78. Hakkarainen M, Gröning M, Albertsson A-C (2003) J Appl Polym Sci 89:867
79. Pannkoke K, Oethe M, Busse J (1998) Cryogenics 38:155
80. Gröning M, Hakkarainen M, Albertsson A-C (2004) Polym Polym Comp 12:491
81. Gröning M, Hakkarainen M, Albertsson A-C (2004) Polym Polym Comp 12:501

Adv Polym Sci (2008) 211: 85–116
DOI 10.1007/12_2007_128

Published online: 2 February 2008

Degradation Products of Aliphatic and Aliphatic–Aromatic Polyesters

Minna Hakkarainen · Ann-Christine Albertsson (✉)

Department of Fibre and Polymer Technology,
School of Chemical Science and Engineering,
The Royal Institute of Technology (KTH), SE-100 44 Stockholm, Sweden
aila@polymer.kth.se

Abstract Analysis of degradation products needs to be included in degradability testing to ensure the environmental adaptability of degradable polymers. Identification of breakdown products from environmental degradation is important for understanding the degradation process, environmental interaction and impact of degradable polymers. With regard to degradability aliphatic polyesters are a particularly interesting group of polymers. They are susceptible to hydrolysis and biological attack due to the ester groups in the main chain. This paper summarizes the work done on the chromatographic analysis of degradation products from the most common aliphatic and aliphatic–aromatic polyesters under different abiotic and biotic conditions including simulated, accelerated and real environmental conditions.

Keywords Chromatography · Degradation product · Degradation · Environmental impact · Polyester

Abbreviations

APCI-MS	Atmospheric pressure chemical ionization-mass spectrometry
CL	Caprolactone
CZE	Capillary zone electrophoresis
ESI-MS	Electrospray ionization-mass spectrometry
DXO	1,5-dioxepan-2-one
GA	Glycolic acid
GC	Gas chromatography
3HB	3-hydroxybutyric acid
HHA	6-hydroxyhexanoic acid
3HV	3-hydroxyvaleric acid
HPA	3-(2-hydroxyethoxy)-propanoic acid
HPLC	High performance liquid chromatography
LA	Lactic acid
MALDI-TOF	Matrix-assisted laser desorption-ionization time-of-flight
MS	Mass spectrometry
PBA	Poly(butylene adipate)
PBG	Poly(butylene glycol)
PBS	Poly(butylene succinate)
PBSA	Poly(butylene adipate-co-succinate)
PBSL	Poly(butylene succinate-co-lactate)
PBT	Poly(butylene terephthalate)
PCL	Polycaprolactone
PDLLA	Poly-DL-lactide
PEA	Poly(ethylene adipate)
PEG	Poly(ethylene glycol)
PES	Poly(ethylene succinate)
PET	Poly(ethylene terephthalate)
PHB	Poly(3-hydroxybutyrate)
PHBV	Poly(3-hydroxybutyrate-co-3-hydroxyvalerate)
PHV	Poly(3-hydroxyvalerate)
PLA	Polylactide
PLG	Poly(lactide-co-glycolide)
PLLA	Poly(L-lactide)
PPT	Poly(propylene terephthalate)
PTMA	Poly(tetramethylene adipate)
PVAl	Poly(vinyl alcohol)
SPE	Solid-phase extraction

1 Introduction

Biodegradable or environmentally degradable polymers are increasingly used as an alternative to traditional commodity plastics in shopping bags, packaging and other single-use applications as well as mulch films for agriculture. The development of truly environmentally degradable and environmentally friendly polymers requires better understanding of the correlations between environmental parameters, degradation rate and breakdown mechanisms. Degradable materials need to have controlled service-life during which their properties and performance should not significantly deteriorate. After the service-life they should degrade in a predictable fashion leaving no harmful degradation products or polymer fragments behind. Extensive characterization of structure, properties and performance during simulated service life as well as characterization of the degradation process and degradation products after disposal is needed to ensure this (Fig. 1). Important questions include how to ensure the performance during service-life and how the degradability of materials should be tested under relevant environmental conditions. The degradability testing should also include analysis of the formed degradation products to evaluate the environmental impact of the material. With regard to environmental degradability aliphatic polyesters are a particularly interesting group of polymers [1]. They are susceptible to hydrolysis and biological attack due to the ester group in the main chain. This paper summarizes the work done on the chromatographic analysis of degradation products from the most common aliphatic and aliphatic–aromatic polyesters under different abiotic and biotic conditions including simulated, accelerated and real environmental conditions.

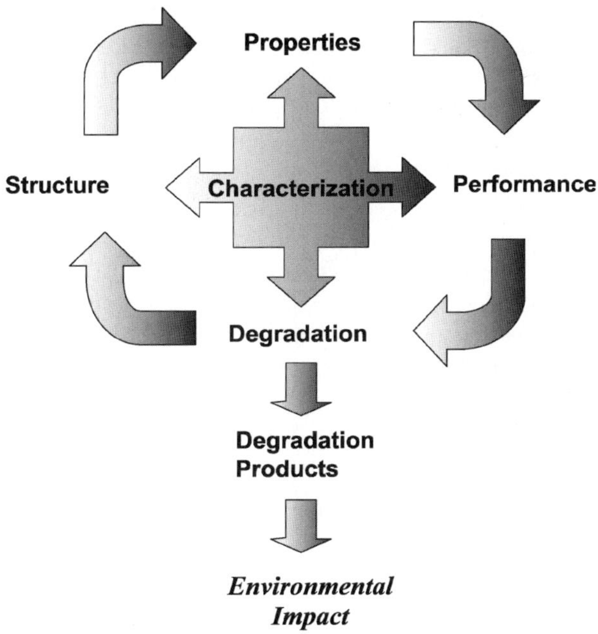

Fig. 1 Characterization of degradable polymers

2 Degradation of Polylactide

Polylactide (PLA) and its copolymers are increasingly used not only in biomedical applications, but also as packaging materials and in other consumer goods. Hydrolysis of poly(L-lactide) (PLLA) and poly-DL-lactide (PDLLA) has been the subject of many studies [2–5]. Hydrolysis rate is influenced by, for example, crystallinity, residual monomer, impurities, molecular weight and molecular architecture [6–9]. The higher susceptibility of the amorphous parts to hydrolysis leads to two-stage degradation of semicrystalline polyesters in aqueous media [10, 11]. The first stage starts with water diffusion into the amorphous regions, which are less organized and allow water to penetrate more easily. The second stage starts when most of the amorphous regions are degraded. The presence of certain enzymes e.g pronase, proteinase K and bromelain increases the hydrolysis rate of polylactide [12]. Proteinase K preferentially degraded PLLA compared to PDLLA [13, 14]. Several actinomycetes from the *Pseudonocardiaceae* family rapidly degraded PLLA [15, 16]. The molecular weight of PLA and PLA oligomers decreased initially by abiotic hydrolysis [17, 18], but later the molecular weight for the samples aged in biotic medium decreased faster than the molecular weight of the samples aged in abiotic medium. Chemical hydrolysis of PDLLA was followed by bio-assimilation of the by-products [19].

The presence of compost microorganisms in the mineral medium significantly accelerated the degradation rate of polylactide compared to the degradation in corresponding sterile mineral medium [20, 21]. After 5 weeks in the biotic environment the films had fragmented to fine powder, while the films in corresponding abiotic medium still looked intact. A rapid molecular weight decrease and increasing polydispersity was observed in the biotic environment. In the abiotic environment only a slight molecular weight decrease was seen and the polydispersity started decreasing towards 2. This indicates different degradation mechanisms, i.e. preferred degradation near the chain ends in the biotic environment and a random hydrolysis of the ester bonds in the abiotic environment. The molecular weight of pure PDLLA films buried in compost was reduced to half of the original value after only 15 days [22]. The rapid molecular weight decrease was explained by the combined effect of thermal degradation, hydrolysis and biodegradation taking place during composting. PLLA disintegrated within 2 weeks in windrow composting [23]. Eco-PLA from Cargill degraded rapidly in bench-scale compost and reached 95% weight loss after 12 days [24]. PLLA was degraded in both aerobic and anaerobic thermophilic conditions, but the degradation rate was highly influenced by temperature [25]. In thermophilic conditions, anaerobic biodegradation of PLLA was even faster than biodegradation under aerobic conditions [26].

2.1
Degradation Products of Polylactide

Solid-phase extraction (SPE) followed by gas chromatographic (GC) analysis showed a complete hydrolysis of poly(lactide-co-glycolide) (PLG) copolymers to water-soluble oligomers and finally to lactic acid (LA) and glycolic acid (GA) in phosphate buffer pH 7.4 at 37 °C or 60 °C [4]. The hydrolysis started by molecular weight decrease, which was followed by weight loss and finally the hydrolysis of water-soluble oligomers to lactic acid and glycolic acid (Fig. 2).

Lactic acid and 3-(2-hydroxyethoxy)-propanoic acid (HPA) were the main hydrolysis products of L-LA, DL-LA and 1,5-dioxepan-2-one (DXO) copoly-

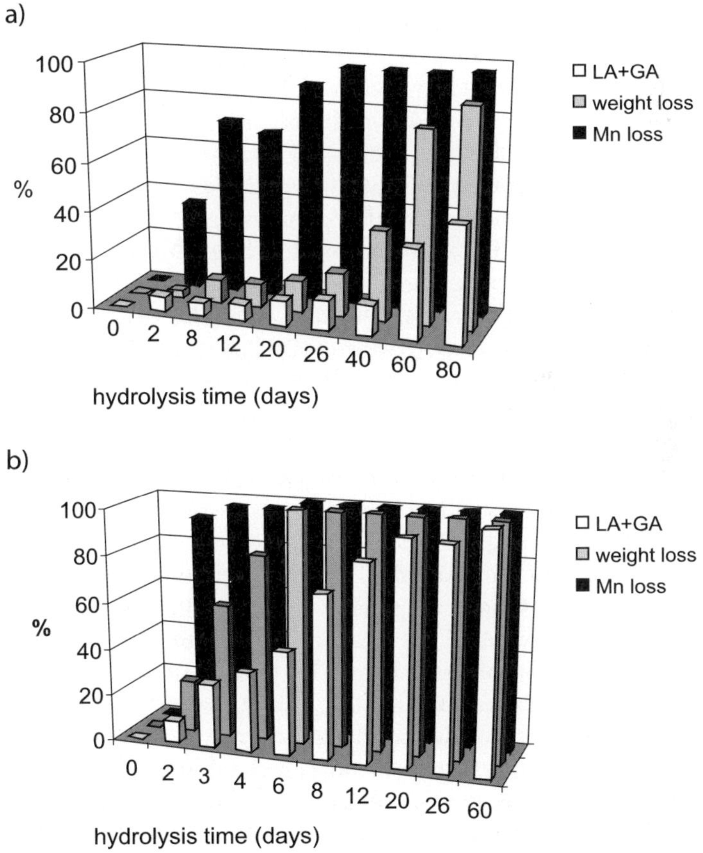

Fig. 2 The loss of number average molecular weight, weight loss and formation of lactic acid and glycolic acid as a percentage of the theoretical amount during hydrolysis of poly(lactide-co-glycolide) (50/50). The degradation starts by molecular weight decrease, which is followed by weight loss and formation of monomeric hydroxyacids

mers at pH 7.4 and 37 °C [27,28]. Gas chromatography-mass spectrometry (GC-MS) analysis of PLLA degradation products showed that product patterns differed depending on whether PLLA was aged in abiotic or biotic mineral medium [20, 21]. As shown in Fig. 3, in the abiotic medium the amount of lactic acid and lactoyl lactic acid increased with aging time due to the hydrolysis of PLLA, while in biotic medium containing compost microorganisms, lactic acid and lactoyl lactic acid formed by abiotic hydrolysis were rapidly assimilated by the microorganisms [20]. New degradation products, i.e. acetic acid, propanoic acid and ethyl ester of lactoyl lactic acid were detected after aging in biotic medium [21]. The concentration of ethyl ester of lactoyl lactic acid increased with aging time, while the fermentation products of lactic acid, i.e. acetic and propanoic acid, were only detected at the beginning of the hydrolysis.

Vert et al. showed that DL- and L-lactic acids and LL-dimer were rapidly assimilated by *F. moniliforme* and *P. putida*, whereas the assimilation of DD-dimer proceeded slowly [17]. A mixed culture of *F. moniliforme* and *P. putida* resulted in faster assimilation of oligomers compared to pure cultures of *F. moniliforme* or *P. putida*. Actinomycete *Kibdelosporangium aridum* degraded 97% of high molecular weight PLLA film within 14 days [29]. Another actinomycete *Saccharothrix waywayandensis* degraded 95% of PLLA films during 4 days in liquid culture containing 0.1% gelatin [30]. Without gelatin only 15% of PLLA was degraded after 7 days. L-lactic acid, the monomeric degradation product first temporarily accumulated but was then totally assimilated by both strains. Vert et al. have also demonstrated the usefulness of capillary zone electrophoresis (CZE) in monitoring water-soluble oligomers formed during hydrolysis of PLA and its copolymers [31]. In vitro aging of lactic acid oligomers showed that degradation did not yield as much monomer as was expected from purely random degradation [32]. Ester scission of larger oligomers formed predominantly dimer. The authors concluded that the ester bond of lactoyl lactic acid is more stable than the ester bonds inside longer oligomer chains.

The composition of the polymer chains, i.e. the content of L-LA, D-LA and/or copolymer units, has a large effect on crystallinity and hydrophilicity and can be utilized to control the hydrolysis rate [33–35]. The half-lives for PLLA, PDLLA and PLG polymers in pH 7.4 phosphate buffer at 37 °C varied from 3 to 110 weeks depending on the amount of L-LA, D-LA and glycolide units in the polymer [36]. Glycolide/caprolactone copolymers with higher C–G bond content (higher degree of randomness) exhibited a higher degradation rate, while sequences with odd numbers of glycolyl units were more resistant to hydrolysis [37]. Li et al. showed that proteinase K could hydrolyze

Fig. 3 The relative amounts of lactide, lactic acid, lactoyl lactic acid and ethyl ester of lactoyl lactic acid formed during hydrolysis of PLLA in biotic and abiotic medium. Reprinted from [20] with permission of Elsevier. © Elsevier (2000) ▶

a)

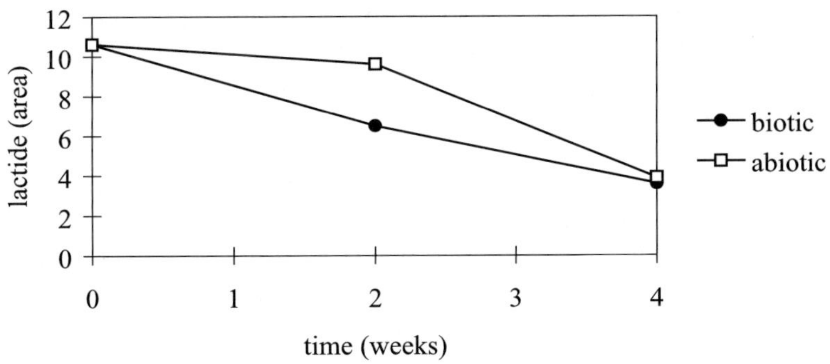
lactide (area)
time (weeks)
biotic
abiotic

b)

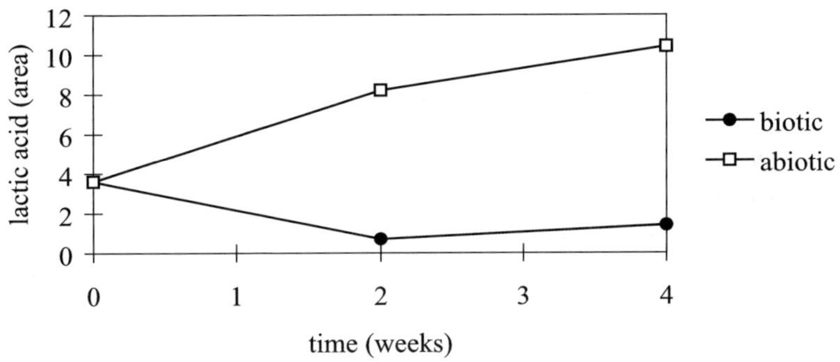
lactic acid (area)
time (weeks)
biotic
abiotic

c)

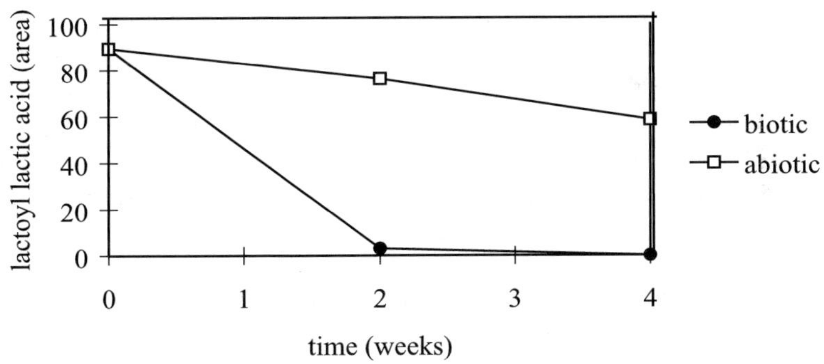
lactoyl lactic acid (area)
time (weeks)
biotic
abiotic

d)

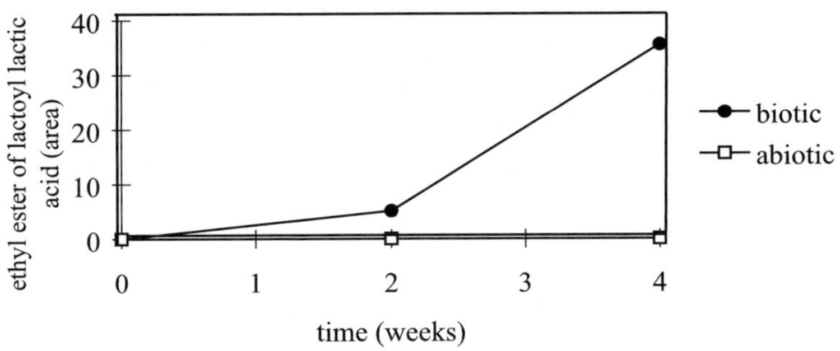
ethyl ester of lactoyl lactic acid (area)
time (weeks)
biotic
abiotic

LL, DL and LD bonds, but DD bonds were not degraded [38, 39]. Pyrolysis of biotically and abiotically aged PLLA showed that the ratio of meso-lactide to L-lactide was lower in samples aged in biotic media compared to samples aged in abiotic media [40]. This also confirms that microorganisms preferentially degrade the L-form of poly(lactide) or its oligomers.

pH of the aging medium affects the hydrolysis rate and mechanism [41]. Hydrolysis proceeded through surface erosion at pH 12 [42], while bulk erosion took place at pH 7.4 [43]. A sequential cleavage near the chain ends took place in alkaline medium, while a random chain scission dominated at acidic pH values [44]. Another study concluded that during alkaline hydrolysis of lactic acid oligomers, lactoyl lactic acid dimer was split off from the oligomeric chains, while hydrolysis at acidic pH formed oligomers with different lengths [45]. The degradation in neutral and alkaline media was explained by intramolecular transesterification (backbiting) through a stable six-membered ring as an intermediate. In the acidic media the preferential cleavage of ester bonds at the terminal hydroxyl end of the oligomer is initiated by protonation of the OH end group, which is followed by formation of an intramolecular hydrogen bridge leading finally to lactic acid being split off. Different product patterns are, thus, expected after aging at different pH.

3 Degradation of Polycaprolactone

Polycaprolactone (PCL) is relatively stable against abiotic hydrolysis, but its susceptibility to biodegradation had already been shown in the 1970s [46, 47]. The proposed degradation mechanism is hydrolysis of the polymer to oligomers and to 6-hydroxyhexanoic acid, an intermediate of ω-oxidation. Biodegradation then proceeds through β-oxidation to acetyl-SCoA, which can undergo further degradation in citric acid cycle. PCL has been shown to biodegrade in pure fungal cultures [48, 49], in compost [50–52], in active sludge [50] and in soil [53]. PCL degrading anaerobic microorganisms are found in several natural environments such as river water, sewage sludge supernatant, farm soil, paddy soil, creek sediment, roadside sand and pond sediment [54, 55]. The enzymes that hydrolyze naturally occurring hydrophobic polyesters such as cutin and lipids may also attack polycaprolactone [56]. The biodegradation rate is controlled by, for example, molecular weight and degree of crystallinity [57]. In an early study Fields et al. found negligible degradation of molecules with molecular weight over 15 000 g/mol [58]. However, later studies have shown that even high molecular weight PCL is biodegraded [57]. The effect of molecular weight on degradability is not well understood and varies with fungal species. Small changes in fungal population may result in dramatic changes in the ability of the consortium to use polymers as a carbon source [59]. Differences in microflora together with the initial morphology of the PCL sample resulted in different mechanisms of erosion [50, 60]. Temperature played an important role in the degradation in compost and anaerobic sludge [59, 61]. The degree of crystallinity increased with degradation indicating preferential degradation of amorphous regions [59]. The interplay between degradation mechanism and the nature of the interaction between microorganisms and polymer substrate have also been studied [62]. If the biofilm was formed on the surface nonpreferential spherical degradation of amorphous and crystalline phases occurred resulting in holes through the film.

Recycling or addition of processing additives slightly decreased the degradation rate compared to the degradation of pure PCL [59]. The presence of starch significantly increased the biodegradation rate of PCL in activated sludge, soil burial and controlled composting [63]. The degradation of PCL/PLLA copolymers [64] and blends [65, 66] have been studied by several authors. *Fusarium solani* and *Fusarium moniliforme* were more effective in degrading copolymers with long PCL sequences [67]. The enzymatic degradation rate of reactive compatibilized blends of poly(lactic acid) and polycaprolactone was faster than the degradation rate of pure PLA or PCL [68]. The degradation rate for physical blends was intermediate between those of PLA and PCL. The degradation of solution-cast PCL/PDLLA blend films by *Pseudomonas* lipase indicated preferential degradation of PCL, while such a preference was not observed for *Candida cylindracea* [69]. Biodegradation of similar blends in soil also indicated preferential degradation of PCL [70]. The in vitro degradation rate of poly(trimethylene carbonate-co-caprolactone) containing 80% CL units was similar to that of PCL [71]. A thin layer of poly(vinylalcohol) (PVAl) on the surface of PCL prevented biodegradation [72].

3.1 Degradation Products of Polycaprolactone

Abiotic hydrolysis of linear, crosslinked and porous PCL resulted in formation of 6-hydroxyhexanoic acid (HHA) and water-soluble oligomers [73]. As seen in Fig. 4 the introduction of crosslinks considerably increased the hydrolysis rate and formation of monomeric 6-hydroxyhexanoic acid. This was mainly explained by the lower degree of crystallinity for the crosslinked PCL homopolymer.

PCL was also degraded to 6-hydroxyhexanoic acid during enzymatic hydrolysis by Lipase Asahi derived from *Chromobacterium viscosum* and lipase F derived from *Rhizopus niveus* [74]. In another study formation of oligomers during biotic hydrolysis of PCL was shown by matrix-assisted laser desorption/ionization time-of-flight mass spectrometry (MALDI-TOF) [50]. Enzymatic degradation of copolymers of 3-hydroxybutyric acid (3HB) and

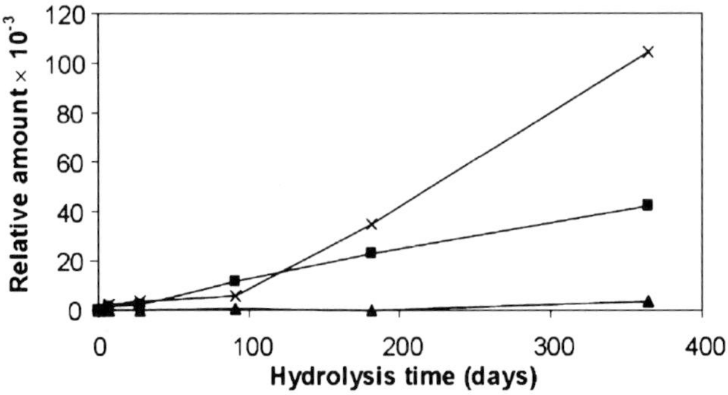

Fig. 4 Migration of 6-hydroxyhexanoic acid from different polycaprolactone homopolymers during hydrolysis in phosphate buffer (▲) porous structure, (■) linear disc and (×) network. Reprinted from [73] with permission of Taylor & Francis. © Taylor & Francis (2007)

6-hydroxyhexanoic acid resulted in the formation of several dimers, trimers and monomers [75]. Different product patterns were obtained when different enzymes were used. After hydrolysis by PHB depolymerase 3HB monomer, 3HB–3HB dimer, HHA–3HB dimer and 3HB–HHA–3HB trimer were detected, but no HHA monomer, 3HB–HHA dimer or HHA–HHA dimer was detected. This indicates that PHB depolymerase from *A. faecalis* was incapable of hydrolyzing the ester bonds of HHA–3HB and HHA–HHA in dimers and trimers. However, after hydrolysis by lipase from *R. delemar* only HHA and HHA–HHA were detected indicating that the lipase only hydrolyzes the ester bonds of HHA–HHA sequences.

A mixed culture of compost microorganisms rapidly assimilated low molecular weight compounds such as 6-hydroxyhexanoic acid and caprolactone, Fig. 5 [60]. Succinic acid, butanoic acid, pentanoic acid and hexanoic acid were identified after biotic hydrolysis of PCL at 50 °C by *Aspergillus* sp. [76]. Cyclic monomer ε-caprolactone was detected as a degradation product after hydrolysis with *Paecilomyces lilacinus* [48]. These products did not accumulate at longer aging times, indicating further assimilation. Degradation of tritium-labeled PCL in medium inoculated with aliquots of sewage sludge showed that after 72 days at 37 °C PCL was completely biodegraded [77]. Approximately 80–90% of the radioactivity initially present in solid PCL was recovered as tritiated water, the rest was incorporated in the biomass.

Fig. 5 GC-MS chromatograms of the low molecular weight products extracted from polycaprolactone films: (**a**) unaged film; (**b**) after 2 weeks in abiotic medium; and (**c**) after 2 weeks in biotic medium. The identity of numbered peaks is 1 = caprolactone; 2 = 6-hydroxyhexanoic acid; 4 = cyclic dimer; 7 = cyclic trimer and 3,5,6 = phthalates. Reprinted from [60] with permission of Wiley–VCH Verlag GmbH & Co. © Wiley–VCH Verlag Gmbh & Co (2002)

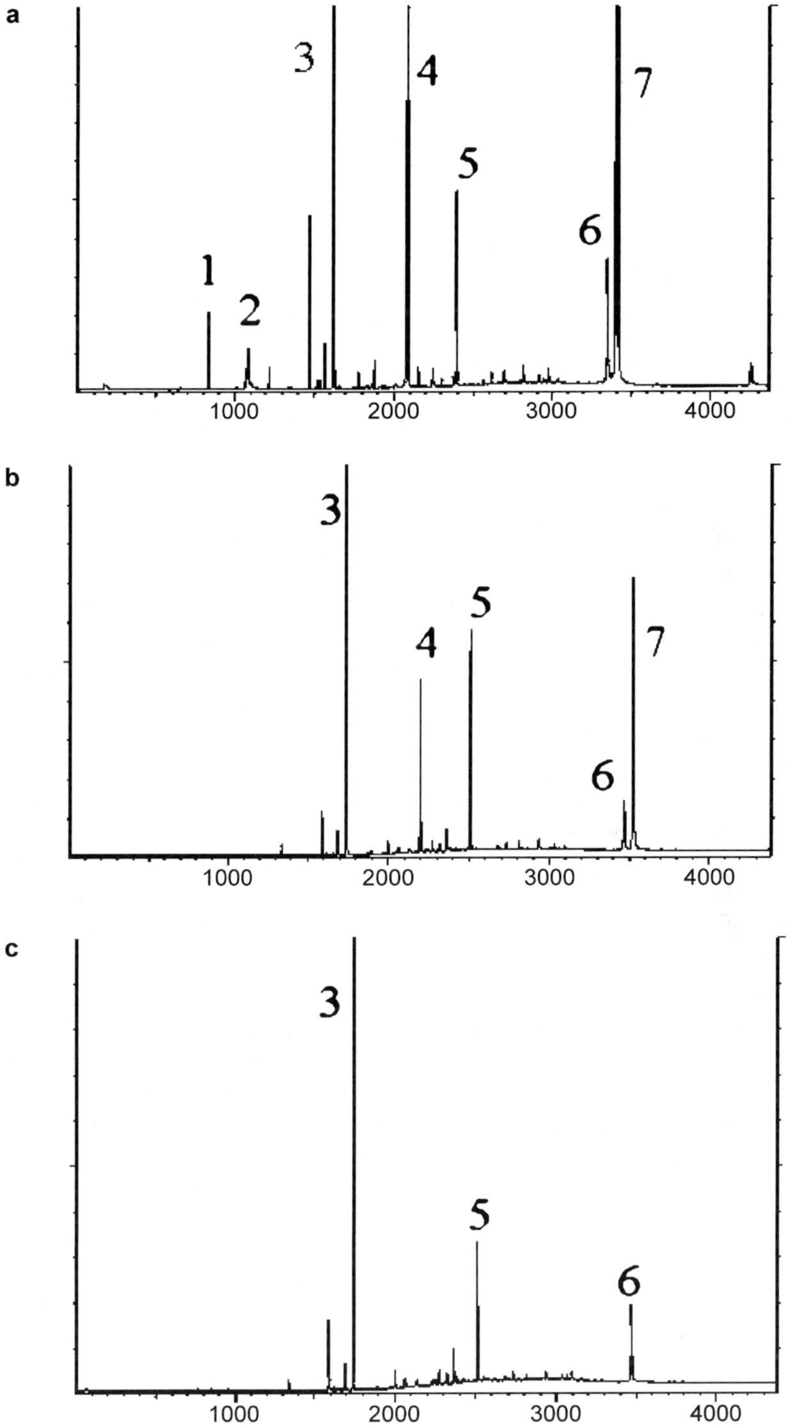
a
3
4
7
5
6
1
2
1000
2000
3000
4000
b
3
5
4
7
6
1000
2000
3000
4000
c
3
5
6
1000
2000
3000
4000

4 Degradation of Poly(3-hydroxybutyrate) and Its Copolymers

Poly(3-hydroxybutyrate) (PHB) and its copolymer poly(3-hydroxybutyrate-co-3-hydroxyvalerate) (PHBV) are rather resistant towards moisture and their chemical hydrolysis at neutral pH and ambient temperature proceeds slowly through random chain scission [78, 79]. The rate of chemical hydrolysis decreases with increasing crystallinity [80]. There is no agreed explanation to how the copolymer composition affects the hydrolysis rate. It has been suggested that it is the crystallinity, rather than the composition that affects the hydrolysis rate. On the other hand, PHBVs of the same crystallinity, but different compositions (45–71 mol% HV), showed decreasing hydrolysis rates with increasing HV-content [81].

Poly(hydroxyalkanoates) are rapidly biodegraded by a wide range of microorganisms [82]. Biodegradation takes place both under aerobic [83] and anaerobic conditions [84, 85] and several PHB depolymerases have been purified from microorganisms [86]. The rate and extent of enzymatic hydrolysis is influenced by copolymer content [81], sequence distribution [87], stereochemistry [88–90] and type of substituents [75]. The introduction of HV units into the PHB chain generally reduces the extent of enzymatic degradation [81, 91] and the enzymatic hydrolysis rate is further reduced by longer side chains at β-carbon [75]. However, at some copolymer compositions the decrease in degree of crystallinity due to copolymerization can also enhance the enzymatic degradation rate. Bacterially produced PHB is an optically active poly[(*R*)-3-hydroxybutyrate]. Chemosynthetic poly(3-hydroxybutyrate) containing both (*R*)- and (*S*)-3HB is also hydrolyzed by PHB depolymerases, but the amount of (*R*)-3HB and (*S*)-3HB units in the chain has a large influence on the hydrolysis rate [88–90]. As an example the enzymatic degradation rate of PHB polymer containing 77% (*R*)-3HB units was several times higher than the degradation rate of microbial poly[(*R*)-3-hydroxybutyrate), while a PHB polymer with 94% (*S*)-3HB units was hardly hydrolyzed at all by a depolymerase from *P. funiculosum* [89]. These results indicate that the PHB depolymerase was not able to hydrolyze the bonds between two (*S*)-3HB units. However, the introduction of some (*S*)-3HB units in the chain decreased the crystallinity, which in turn enhanced the degradation rate at moderate (*S*)-3HB contents.

PHB depolymerase can hydrolyze only polymer chains in the surface layer of the films [81]. The enzymatic hydrolysis of PHB and poly(3-hydroxyvalerate) (PHV) is, thus, a heterogeneous erosion process proceeding from the surface [92]. The effect of solid-state structure on enzymatic degradability has been studied by analysis of films with various degrees of crystallinity, lamella crystal sizes and spherulite sizes [93]. The results demonstrated that the crystallinity and lamellar crystal size play an important role in the degradation process. A lower degree of crystallinity or thinner lamella crystals enhances the enzymatic degradation rate. During initial stages the enzymes preferentially attack the amorphous phase of PHB [94, 95]. Later, however, both amorphous and crystalline regions are degraded without preference. PHB and PHBV degrade in several natural environments, such as soil, compost and seawater [96–98]. Degradation is mainly characterized by large weight loss, but also to some extent by molecular weight decrease [98]. During composting of organic material, moisture is present and the temperature may reach 60–80 °C. The effect of abiotic factors, such as water, air and temperature, on the degradation of PHBV in garden waste compost has been investigated using simulated and natural environments [99, 100] including exposure to sterile water or air at 60 °C, pure fungal cultures and garden waste compost. The results indicated that the degradation in compost was mainly due to microbial action.

4.1
Degradation Products of Poly(3-hydroxybutyrate)

Several studies have shown that enzymatic hydrolysis of PHB and PHBV produces water-soluble monomers and oligomers as degradation products. The type of PHB depolymerase, copolymer composition and stereochemistry all influence the product pattern. Water-soluble degradation products, mainly monomers and dimers were formed during the enzymatic degradation of natural PHB and PHBV [101, 102]. The dimer of 3-hydroxybutyric acid was the primary product of enzymatic hydrolysis of poly(3-hydroxybutyrate) film [102]. However, on prolonged aging it was further hydrolyzed to monomer. Stereochemistry of the PHB polymer influenced the degradation product composition [103]. The enzymatic hydrolysis products of bacterial poly[(*R*)-3-hydroxybutyrate] were monomer and dimer, but the stereoirregular poly(3-hydroxybutyrate) produced monomer, dimer, trimer and tetramer as degradation products. This suggests that PHB depolymerase was incapable of hydrolyzing the ester bonds in (*S*)-3HB units. The enzymatic degradation of poly(3-hydroxybutyrate-co-6-hydroxyhexanoate) by a depolymerase from *A. faecalis*, resulted in the formation of 3-hydroxybutyric acid monomer, 3-hydroxybutyrate-3-hydrohybutyrate dimer, HHA–3HB dimer and 3HB–HHA–3HB trimer [75]. 6-Hydroxyhexanoic acid monomer or dimer were not detected indicating that the PHB depolymerase was incapable of hydrolyzing the ester bonds between HHA–3HB and HHA–HHA. Enzymatic hydrolysis of racemic PHB samples produced a mixture of monomer, trimer, trimer and tetramer of 3-hydroxybutyric acid, while syndiotactic PHB films were hardly hydrolyzed by PHB depolymerase from *A. faecalis* [90]. Almost no adsorption of enzyme took place on the surface of syndiotactic films indicating that the binding domain of PHB depolymerase has no affinity towards syndiotactic structure.

The enzymatic hydrolysis of PHB by PHA depolymerase from *Acidovorax* sp. TP4 produced 3HB monomer as a major product, while mainly 3HB dimer was produced by PHA depolymerase from *Ralstonia pikettii* T1 [104]. The enzymatic hydrolysis of chemosynthesized atactic PHB by the same PHA depolymerase from *Acidovorax* sp. resulted in the formation of monomer, dimer and trimer [105]. This showed that chemosynthesized atactic PHB could be degraded by a natural PHA depolymerase. Compared to the enzymatic hydrolysis of bacterial PHB the hydrolysis product pattern contained higher molecular weight products. 3-Hydroxybutyric acid and its dimer were identified by high-performance liquid chromatography (HPLC) as degradation products from PHBV and its blends with atactic PHB after enzymatic hydrolysis by *Pseudomonas lemoignei* [106]. Higher oligomers up to heptamer were detected by atmospheric pressure chemical ionization-mass spectrometry (APCI-MS) and electrospray ionization-mass spectrometry (ESI-MS). It was also shown that water-soluble oligomers of atactic PHB are assimilated by selected bacterial strains [107]. Linear and cyclic 3HB and 3HV were hydrolyzed by depolymerases isolated from *Aspergillus fumigatus* and *Alcaligenes faecalis* [108].

3-Hydroxybutyric acid, 3-hydroxyvaleric acid, 3-hydroxybutyrate dimer, 3-hydroxybutyrate-3-hydroxyvalerate dimer and 3-hydroxyvalerate dimer were identified as PHBV degradation products after 10 days with *Aspergillus fumigatus* at 25 °C [99]. After 21 days the polymer was completely degraded to water-soluble products. When biotic hydrolysis was continued 3-hydroxybutyric acid and 3-hydroxyvaleric acid disappeared and acetic, butyric and valeric acid were detected instead. At the same time, the mineral medium changed from transparent to yellow due to the excretion of these metabolites. In the sterile control no degradation products were detected during the same period. Acetic acid, propanoic acid, butyric acid and valeric acid were also detected when PHBV was fermented by a selective culture in mineral salt medium [85]. The biodegradation of poly[(*R*,*S*)-3-hydroxybutyrate], atactic PHB, and its blends with natural PHB and PLLA was investigated in soil [109]. No signals related to PHB or PLLA were detected in the ESI-MS spectra of the methanol/chloroform extracts of post-degradation soil. Neither were any ecotoxicological effects detected by using the terrestrial plant growth test.

During aging of PHVB in sterile water at pH 7 and 60 °C, 2-butenoic acid (crotonic acid), 2-pentenoic acid, 3-hydroxybutyric acid, 3-hydroxyvaleric acid, 3-hydroxybutyrate dimer, 3-hydroxybutyrate-3-hydroxyvalerate dimer and 3-hydroxyvalerate dimer were formed [99]. The weight loss was, however, only 2% after 200 days at 60 °C. Monomers, oligomers and derivatives, produced by dehydration at the OH-terminus were identified after alkaline hydrolysis of PHB [110]. In accordance CZE showed that the accelerated hydrolysis of PHB leads to the formation of hydroxyacid oligomers and a series of peaks formed by a side reaction leading to a C=C bond at the noncarboxylic acid end [111]. Kinetics of the abiotic hydrolysis of PHB in acid and alkaline media were monitored by following the formation of two monomeric hydrolysis products, 3-hydroxybutyric acid and crotonic acid [112]. The monomeric products were the main hydrolysis products after hydrolysis in alkaline solution, but were not released in acidic solution. The release of monomeric products was 30-times faster from the amorphous PHB granules compared to crystallized PHB precipitates and solvent-cast films.

5
Degradation of Poly(alkylene dicarboxylate)s

The abiotic hydrolysis of poly(alkylene dicarboxylate)s generally proceeds slowly. The hydrolysis rate of poly(butylene adipate) (PBA) was controlled by crystallinity and molecular weight [113]. The hydrolysis of poly(butylene succinate) (PBS) and PBA in water and phosphate buffer pH 7.0 showed that PBS was hydrolyzed somewhat faster than PBA with corresponding molecular weight [114]. Lower molecular weight or branching enhanced the hydrolysis rate and hydrolysis proceeded faster in water compared to phosphate buffer. In accordance hydrolytic degradation of PBS, PBA and poly(butylene adipate-co-butylene succinate) (PBSA) copolymers in ammonium chloride buffer solution pH 10.6, showed that the degradation rate decreased when the adipate content in the copolymer increased [115]. The hydrolysis rate of PBS in alkaline solution was higher for isothermally crystallized samples compared to melt-quenched samples with a similar degree of crystallinity [116]. This was explained by the differences in the internal morphology of the spherulites: in an isothermally crystallized sample the spherulite consists of coarse and loosely packed fibrils, whereas a melt-quenched sample contains finer and tightly packed fibrils. Copolymerization affected the hydrolysis rate and mechanism [117]. The in vitro hydrolysis of block copolyester poly(ethylene-succinate)-*b*-poly(butylene glycol) (PES/PBG) proceeded slowly mainly through surface erosion [118], while poly(ethylene succinate)-*b*-poly(ethylene glycol) (PES/PEG) [119] degraded rather rapidly through hydrolysis of both bulk and surface.

An early study showed that low molecular weight PBS, PBA and many related polyesters are enzymatically hydrolyzed by lipases such as *Rhizopus delemar* [120]. Generally, the enzymatic hydrolysis rate is highly influenced by crystallinity and chemical structure. Fields et al. studied the biodegradability of polyesters made from C_2–C_{12} diacids coupled with C_4–C_{12} dialcohols and found that the distance between ester groups was an important factor affecting the susceptibility to biodegradation [121]. The fastest enzymatic degradation occurred for polyesters made of diols containing 4 or 6 carbons and diacids containing 6–10 carbons. Several studies have shown significant increase in enzymatic hydrolysis rate after copolymerization of PBS with longer dicarboxylic acids or with, for example, ethylene glycol [122–

124]. Copolymerization both decreased the crystallinity of PBS and introduced groups that are more susceptible to enzymatic attack. In accordance enzymatic hydrolysis of different poly(butylene succinate-co-butylene adipate) copolyesters by *Candida cylindracea* showed that the hydrolysis rate was strongly influenced by both the degree of crystallinity and the chemical structure [124]. The enzymatic degradation rate also depended on the substrate specificity of the selected enzymes [125]. PBS-degrading mesophilic and thermophilic microorganisms have also been studied [126, 127]. Poly(butylene succinate), poly(butylene succinate-co-adipate) and poly(ethylene succinate) were degraded by the strain TB-13 isolated from soil samples [128]. However, the isolated strain could not utilize these plastics as a sole carbon source and it could not utilize succinic acid or adipic acid. A thermophilic strain, identified as *Bacillus stearothermophilus*, capable of degrading poly(butylene succinate-co-butylene adipate) was isolated from 95 soil samples obtained from different locations [129].

PBS and PBA polyesters are degradable in compost, in moist soil, in fresh water with activated sludge and in seawater. Biodegradability in activated sludge and in simulated landfill tests generally increased as the butylene adipate content in poly(butylene succinate-co-butylene adipate) copolyesters increased [115]. However, the PBSA 40/60 degraded faster than the PBSA 20/80, which was explained by the lower melting point and lower crystallinity of PBSA 40/60. The degradation rate for aliphatic polyesters based on different diols and dicarboxylic acids under aerobic conditions in river water at 25 °C was highly dependent on the chemical structure [55]. PES degraded rapidly and 100% weight loss had occurred after 14 days. Weight losses for poly(ethylene adipate) (PEA), PBA and PBS were 95%, 21% and 1% after 28 days, respectively. In addition to the chemical structure of the polyester the type of natural water had a large influence on the degradation rate [130].

5.1 Degradation Products of Poly(alkylene dicarboxylate)s

The most-abundant hydrolysis products identified by GC-MS after abiotic hydrolysis of PBS and PBA were the different monomers, i.e. 1,4-butanediol, succinic acid and adipic acid [131]. In addition small amounts of dimers and trimers were also detected. Comparison of weight loss and the amount of monomeric hydrolysis products showed secondary hydrolysis of water-soluble oligomers to monomeric products, Fig. 6. Depending on the type of bacterial lipase complete hydrolysis of PBSA to water-soluble products took from 3 to 24 h [132]. Approximately 80% of hydrolysis products were oligomers containing one or two ester-linkages, but oligomers containing up to six ester-linkages were detected. Only small amounts of monomeric hydrolysis products were formed. Oligomers containing one ester-linkage

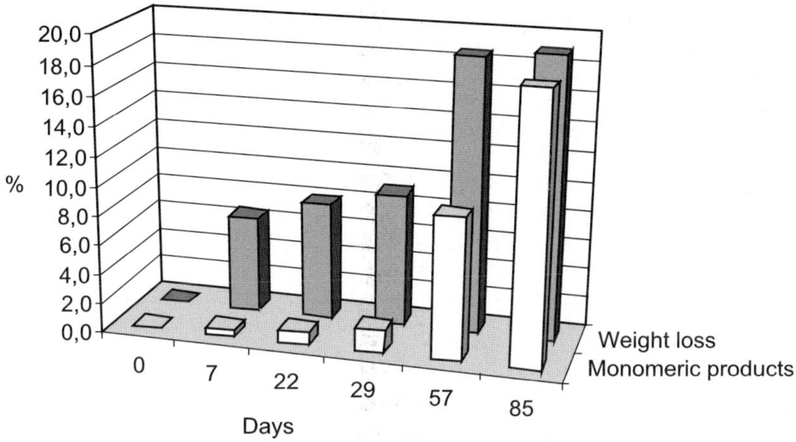

Fig. 6 Comparison of weight loss and the amount of monomeric hydrolysis products, i.e. 1,4-butanedion and succinic acid, resulting from the hydrolysis of poly(butylene succinate) shows further hydrolysis of oligomeric hydrolysis products. Reprinted from [131] with permission of Elsevier. © Elsevier (2004)

accumulated during the course of hydrolysis, while longer oligomers were further hydrolyzed by the lipase. The oligomers with adipate as an end-group were hydrolyzed more rapidly than oligomers without adipate. Twenty-eight different water-soluble monomers and oligomers were identified after hydrolysis of commercial PBSA (Bionolle 3020) by *Chromobacterium* extracellular lipase [133]. Twenty of the species were polyester-based monomers and oligomers, while eight species were diurethane compounds. The polyester-based oligomers were rapidly further biodegraded, while biodegradation of diurethane oligomers proceeded considerably slower. PBSA and PBS were mainly degraded to dimers during enzymatic hydrolysis by Lipase Asahi derived from *Chromobacterium viscosum* and lipase F derived from *Rhizopus niveus* [74]. Enzymatic hydrolysis of PEA, PES, PBS, PBA and PBSA by *Aspergillus fumigatus* initially produced water-soluble oligomers and monomers [134]. According to this study PHB and PHV degraded exclusively to monomer (85–90%) and dimer (10–15%), while PEA, PES and poly(tetramethylene adipate) (PTMA) mostly degraded to monoglycol esteracids. Polymer chain structure strongly influenced the activity of the hydrolase toward specific polymers in a series of closely related polyesters.

Enzymatic hydrolysis of PBS and poly(butylene succinate-co-lactate) (PBSL) by lipase *Pseudomonas cepacia* gave 4-hydroxybutyl succinate as the main hydrolysis product with traces of succinic acid and 1,4-butanediol as well as lactic acid in the case of PBSL [135]. In addition, the hydrolysis rate of the carboxyl end-capped PBS was much slower than that of the original or hydroxyl end-capped PBS indicating preferential *exo*-type chain scission from the carboxyl terminus. Poly(butylene succinate-co-butylene sebacate) and poly(butylene succinate-co-butylene adipate) were enzymatically hydrolyzed by Lipase from *Mucor miehei* or from *Rhizopus arrhizus* [136]. HPLC/ESI-MS analysis showed a preferential cleavage of sebacic bonds, whereas succinic ester bonds were hydrolyzed faster than adipic ester bonds. Aliphatic poly(butylene succinate-co-cyclic carbonate) was degraded to monomeric and oligomeric products by lipase B *Candida Antartic* and lipase *Porcine Pancreas* [137]. Degradation tests of ^{14}C-labeled poly(butylene adipate) showed that microorganisms were able to further degrade oligomers formed during the hydrolysis of PBA [113].

6
Degradation of Aliphatic–Aromatic Polyesters

The hydrolysis of aromatic polyesters like poly(ethylene terephthalate) (PET) and poly(butylene terephthalate) (PBT) at room temperature or under environmental conditions is an extremely slow reaction and no significant direct microbial or enzymatic attack on pure PET or PBT have been observed [138]. Natural environmental processes are, thus, unable to break down products made of PET and they maintain their strength and integrity for a long period of time [139, 140]. Blown PET bottles exposed to 45% relative humidity, wet soil and 100% relative humidity at 20 °C have life expectancies of 48, 35 and 27 years, respectively [139]. The life expectancies at relative humidities lower than 45% are considerably longer than that.

The incorporation of small amounts of hydrolyzable aliphatic ester linkages in the backbone of aromatic polyesters modifies their hydrolytic properties without significantly altering the good physical and mechanical properties. Early investigations on the biodegradation of aliphatic–aromatic copolyesters came to the conclusion that significant degradation occurred only when the aromatic content of the copolyester was relatively low [141]. Witt el al. first reported on a significant biological attack on a copolyester poly(trimethylene decanoate-co-trimethylene terephthalate) with 50 mol % terephthalic acid in the acid component [142]. The block copolymers were more susceptible to microbial attack than the random copolymers, indicating preferential degradation of longer aliphatic sequences. Copolyesters of PET, poly(propylene terephthalate) (PPT) and poly(butylene terephthalate) (PBT) with adipic acid and sebacic acid units were degraded in a compost simulation test at 60 °C and in soil at room temperature [143]. The adipic acid-based copolymers degraded considerably faster in the compost (100% weight loss after 8 weeks) compared to sebacic acid copolyesters (30% weight loss after 8 weeks). The weight loss decreased with increasing terephthalic acid content and the weight loss in soil was significantly lower than the weight loss in compost. Acceptable user properties were combined with biodegradability in the composition range 30–55 mol % terephthalic acid [144, 145].

The biodegradation rate of aliphatic–aromatic copolyesters by mixed culture of compost microorganisms depended on the terephthalic acid content [144]. Above 60 mol % terephthalic acid the degradation rate became so slow that the material was no longer suitable for composting. Respirometric tests of the copolyester Ecoflex in mature compost matrix, where the CO_2 evolved during the metabolic conversion was determined, showed that more than 90% of the material was metabolized [146]. Degradation of poly(butylene adipate-co-terephthalate) has also been studied in distilled water at 70 °C, acidic conditions at pH 2.3 and in enzymatic media containing *Pseudomonas cepacia* or *Candida cylindracea* [147]. Both the hydrolytic and enzymatic degradation rate decreased with increasing terephthalic acid content in the copolyester. Over 20 polyester degrading species were isolated by using compost as a microbial source [148]. Two actinomycete strains, identified as *Thermomonospora fusca* strains, exhibited about 20-fold higher polyester degradation rates than usually observed in a common compost test. However, *Thermomonospora fusca* was not able to metabolize significant amounts of monomers generated by polymer hydrolysis [149] and the excretion of the hydrolase was inhibited by the monomeric degradation products from the polyester [150].

Blends and copolymers of PET and PBT with several easily hydrolyzable aliphatic polyesters including polycaprolactone [151], poly(ethylene adipate) (PEA) [152], poly(glycolide) [153] and poly(L-lactide) [154, 155] have been prepared to enhance the environmental degradation rate. Degradation of PCL/PET blends have been studied under different environmental conditions including full-scale composting, soil burial and bench-scale accelerated aerobic degradation [151]. The biodegradation of the blend samples was well below the values expected from the behavior of individual homopolymers under the same environmental conditions. Only very limited hydrolysis was detected when PCL/PET copolymers were subjected to *Rhizopus delemar* and the susceptibility to hydrolysis decreased with increasing aromatic content [141].

6.1 Degradation Products of Aliphatic–Aromatic Polyesters

Water-soluble aromatic oligomers with one or two terephthalic acid units and all monomers, i.e. terephthalic acid, ethylene glycol, 1,3-propanediol and 1,4-butanediol, were rapidly metabolized by microorganisms in aqueous systems, in soil and in compost at 60 °C [156]. The final degradation of longer aromatic oligomers took a significantly longer time. In compost at elevated temperatures also longer oligomers disappeared, probably due to chemical hydrolysis. The polymer chain of Ecoflex copolyester was cleaved within a few days by isolated actinomycetes and at the end of the test only water-soluble intermediates were found in the aqueous test system [146].

The isolated microorganisms were not able to metabolize the monomers and GC-MS analysis showed that the monomers accumulated in aqueous media. In another study aromatic as well as aliphatic oligomers were detected as degradation intermediates from the copolyesters [149]. On prolonged aging these oligomers were further degraded to monomers. A subsequent complete metabolization of the monomers took place if a mixed culture of microorganisms from a compost was added. No accumulation of aromatic oligomers was observed during the composting process. Enzymatic hydrolysis of both aliphatic and aromatic polyesters by lipase from *Pseudomonas species* was shown [157]. However, compared to the cleavage rates of aromatic model esters, the cleavage rates of aliphatic model esters were larger by more than an order of magnitude. Degradation products of poly(butylene succinate-co-terephthalate) after enzymatic hydrolysis by Lipase from *Pseudomonas cepacia* were analyzed by LC-MS [158]. Water-soluble oligomers up to hexamer were detected after the hydrolysis. The oligomeric fragments were slowly hydrolyzed by secondary hydrolysis into 4-hydroxyl succinate and terephthalate. On the basis of the obtained data an *endo*-hydrolysis mechanism was proposed.

7
The Effect of Copolymer Composition on Degradation Products

Copolymerization has been extensively used to modify the hydrolysis rate. In a recent study we also showed controllable degradation product migration through predetermined alterations in copolymer composition [159]. Degradation rate and formation of hydrolysis products were tuned by copolymerizing caprolactone with different amounts of hydrophilic 1,5-dioxepan-2-one. As seen from Fig. 7 the total release rate of monomeric degradation products as determined by GC-MS (Fig. 7a) and water-soluble degradation products as estimated from the mass loss (Fig. 7b) increased gradually as the DXO content in the homo- and copolymers was increased from 0 to 100%. The CL-content in each polymer is given in its name, for example CL80 consists of 80% CL-units and 20% DXO-units.

Figure 8 shows the amount of 3-(2-hydroxyethoxy)-propanoic acid (HPA) and 6-hydroxyhexanoic acid (HHA) migrating from the CL/DXO copolymers as a function of copolymer composition and hydrolysis time. Figure 8a shows that the amount of HPA released from the materials increased both as a function of hydrolysis time and as the DXO-content in the copolymer increased. In the case of HHA release the situation was more complicated and the largest amount of HHA was released from the copolymers with intermediate CL-contents (Fig. 8b). In relation to the CL-content, the largest amount of HHA was released from the DXO-rich copolymers (Fig. 8c). This was explained by the faster hydrolysis rate of DXO-units, which also enhances the release of HHA.

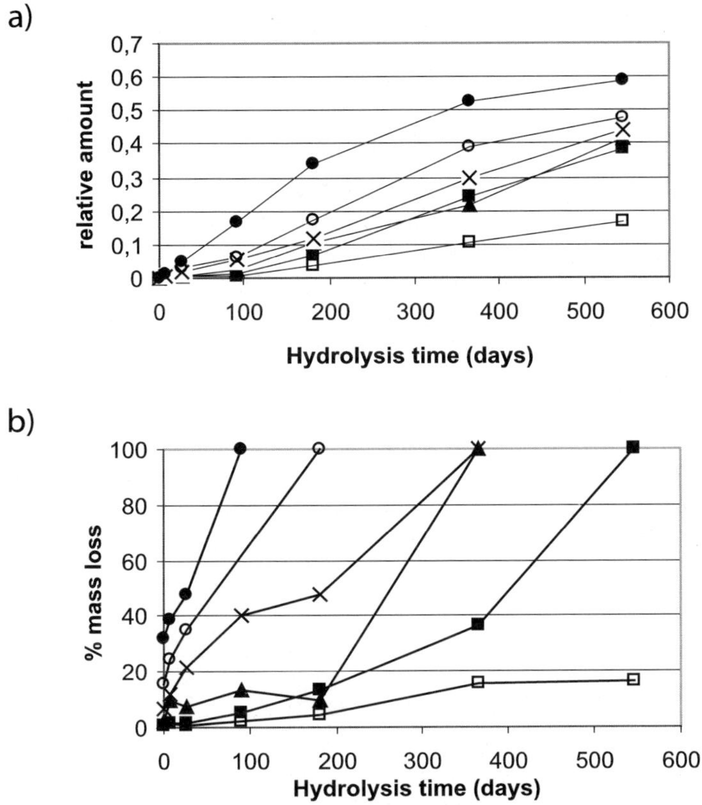

Fig. 7 **a** Relative amount of monomeric hydroxy acids migrating from the crosslinked CL/DXO copolymers. (**b**) Mass loss of the crosslinked CL/DXO copolymers. (□) CL100, (■) CL80, (▲) CL60, (×) CL40, (○) CL20 and (•) CL0 during the hydrolysis. The name of the material indicates the caprolactone content in the polymer, e.g. CL80 has 80 mol % caprolactone units and 20 mol % DXO-units. Reprinted from [159] with permission of American Chemical Society. © American Chemical Society (2007)

8 The Effect of Macromolecular Architecture on Degradation Products

Macromolecular engineering is a promising tool for tailoring the degradation rate of degradable polyesters. We have shown that macromolecular architecture also has a large influence on the degradation product patterns of aliphatic polyesters [160]. The molecular weight changes were similar for CL/DXO triblock and multiblock copolymers, however, large differences were observed in the release rate of acidic degradation products as well as in the product patterns, Fig. 9. After 364 days the total amount of water-soluble monomeric hydroxyacids released from the CL/DXO triblock copolymer was three times higher compared to the CL/DXO multiblock copolymer with

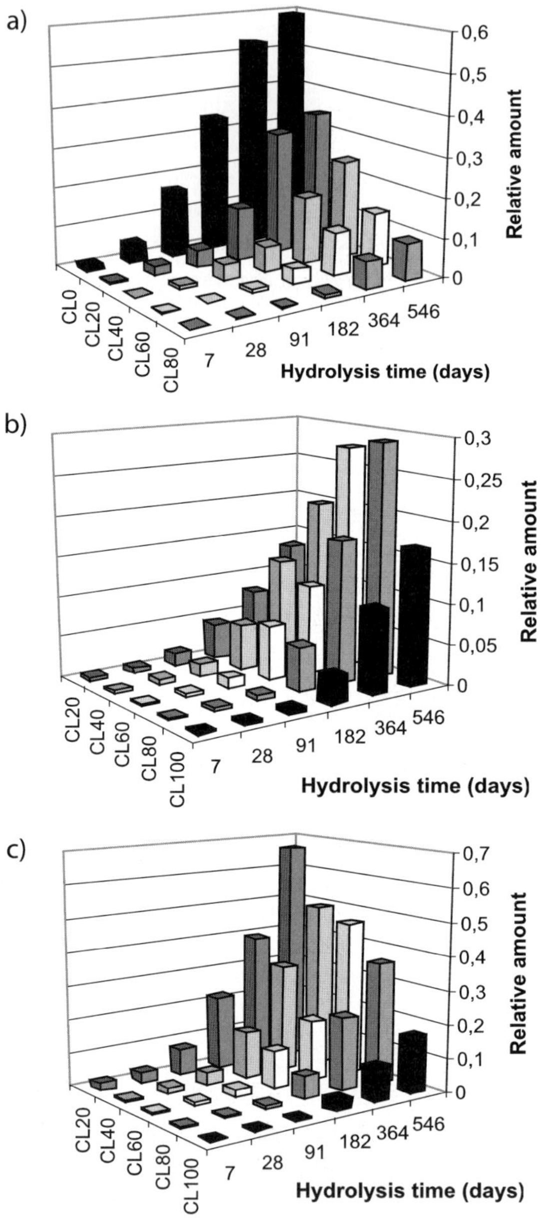

Fig. 8 Relative amount of (**a**) 3-(2-hydroxyethoxy)-propanoic acid (HPA) and (**b**) 6-hydroxyhexanoic acid (HHA) migrating from the cross-linked CL/DXO copolymers. **c** 6-Hydroxyhexanoic acid migrating from the cross-linked CL/DXO copolymers relative to the original CL content in the copolymer. The name of the material indicates the caprolactone content in the polymer, e.g. CL80 has 80 mol % caprolactone units and 20 mol % DXO-units. Reprinted from [159] with permission of American Chemical Society. © American Chemical Society (2007)

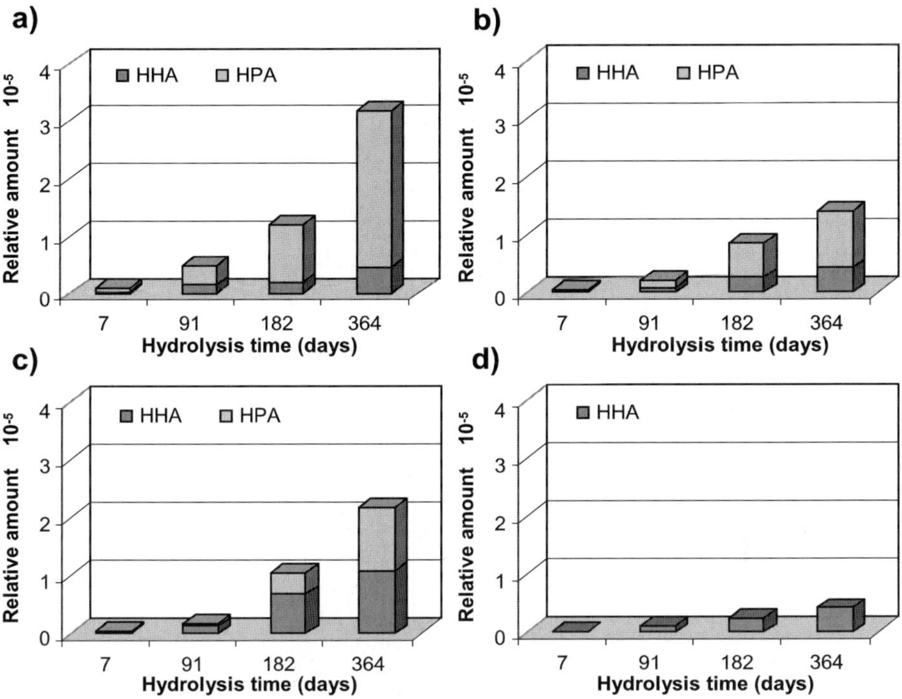

Fig. 9 The relative amount of 6-hydroxyhexanoic acid (HHA) and 3-(2-hydroxyethoxy)-propanoic acid (HPA) formed during hydrolysis of **a** DXO/CL/DXO triblock copolymer. **b** CL/DXO multiblock copolymer. **c** Random crosslinked CL/DXO copolymer and **d** PCL homopolymer. All of the copolymers had 60 mol % CL units and 40 mol % DXO units. The polymers were hydrolyzed for different times in phosphate buffer pH 7.4 and 37 °C. After the predetermined hydrolysis times the monomeric degradation products were extracted by solid-phase extraction and analyzed by GC-MS. Reprinted from [160] with permission of American Chemical Society. © American Chemical Society (2007)

the same copolymer composition (60 mol % CL and 40 mol % DXO). This difference was mainly due to the large amount of HPA formed due to the faster hydrolysis of hydrophilic DXO-blocks in the CL/DXO-triblock copolymer. The largest amount of 6-hydroxyhexanoic acid was formed in the case of random crosslinked CL/DXO copolymer. This was explained by random distribution of the "weak" DXO-linkages and the amorphous nature of the crosslinked polymer.

Figure 10 shows that the by far largest amount of water-soluble oligomers was released from the CL/DXO triblock copolymer. This is probably due to the combination of faster hydrolysis rate and higher water solubility of the DXO-oligomers. Relative to the mass loss the lowest amount of oligomers and largest amount of monomeric hydrolysis products was released from the crosslinked CL/DXO copolymer. A larger amount of chain scissions is needed

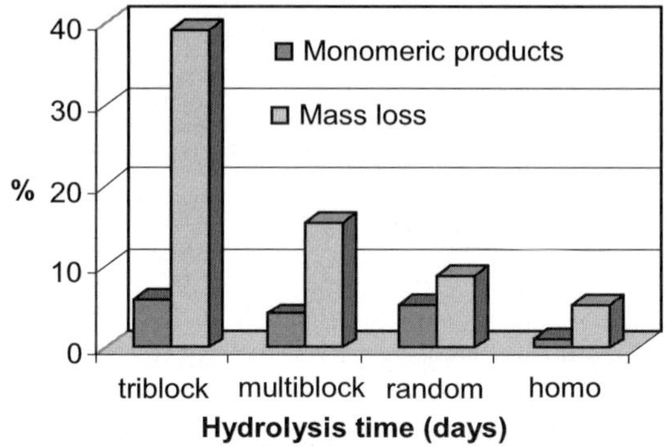

Fig. 10 Comparison of mass loss and relative amount of monomeric degradation products released from DXO/CL/DXO triblock copolymer, CL/DXO multiblock copolymer, random crosslinked CL/DXO copolymer and PCL homopolymer after 182 days of hydrolysis in phosphate buffer pH 7.4 and 37 °C. The monomeric products were extracted from the buffer solution with solid-phase extraction and analyzed by GC-MS. All of the copolymers had 60 mol % CL units and 40 mol % DXO units. Reprinted from [160] with permission of American Chemical Society. © American Chemical Society (2007)

for the formation of water-soluble products from the crosslinked polymer, which pushes the product pattern towards monomeric products.

9 The Role of End Groups

The nature and number of end groups is important in determining the hydrolytic degradation rate and characteristics. The hydroxyl end group played a crucial role in the degradation of PLLA in both alkaline and acidic medium as the protection of the hydroxyl end group substantially retarded the hydrolysis rate [45]. It was shown that the hydrolysis rate of end-capped polylactide-co-glycolide copolymers was 2–3 times lower than the hydrolysis rate of uncapped polymer [161]. Higher water uptake of uncapped polymers might explain this difference [162]. The mass and molecular weight loss of –COOH terminated polylactides was high compared to Cl- and NH_2-terminated polylactides [163]. These end-group effects increased with increasing number of chain-ends. The hydrolysis mechanism of monodisperse oligo(lactic acid)s esterified with *N*-(2-hydroxypropyl)methacrylamide was strongly influenced by the nature of the chain end [164]. The oligomers with free hydroxyl groups degraded predominantly by chain end scission via a backbiting mechanism. However, if the chain ends were protected by acetylation then random chain scission became the rate limiting step. The erosion of PCL has been shown to proceed in the vicinity of chain-ends [51]. In accordance the presence of phthalic end-groups as end-cappers reduced the biodegradability of PCL [165].

Also the carboxyl end groups formed during the hydrolysis of polyesters are capable of catalyzing the hydrolysis of remaining ester bonds, a phenomenon called autocatalysis [166, 167]. In the case of massive specimens of PLA or its copolymer, PLG, the hydrolysis rate is faster in the center than at the surface of the specimen [2]. The hydrolysis products localized near the surface can migrate to the surrounding aging medium, while the concentration of acidic hydrolysis products increases in the center. This higher carboxylic acid concentration catalyzes ester hydrolysis, which results in heterogeneous degradation and multimodal molecular weight distributions as the molecular weight at the center of the specimen decreases more rapidly than the molecular weight at the surface. Finally, hollow structures are formed when the internal material, which is totally transformed to soluble oligomers, dissolves in the aqueous medium. Surface-interior differentiation with faster internal degradation has also been observed for other PLA copolymers [168] and for semicrystalline polyesters [3, 169]. However, in the case of semicrystalline polymers no hollow structures were obtained, because of the crystallization of internal hydrolysis products.

10 Environmental Impact of Degradation Products and Intermediates

A limited amount of work has been published on the ecotoxical impact of degradable polyesters and the environmental toxicity of their degradation products. Lactic acid, alkyl lactate esters and lactate salts generally show favorable environmental characteristics in biodegradability and ecotoxicity tests [170]. Lactic acid and sodium lactate appeared toxic to earthworms at large concentrations, whereas calcium lactate was not [171]. However, the detected vermitoxicity corresponded to amounts that can hardly be found under real compost or soil conditions and the authors concluded that PLA degradation by-products are not toxic to earthworms. No toxicological effects were detected for polylactide or the poly(lactide-urethane) chain linked with 1,4-butane diisocyanate when the ecotoxicological impact during composting was evaluated by flash test, measuring the inhibition of light production of *Vibrio fischeri* and by plant growth tests with cress, radish and barley [172]. However, some toxicity was detected in the case of the poly(lactide-urethane) chain linked with 1,6-hexamethylene diisocyanate. Release of toxic degradation products has been shown even for other poly(ester-urethane)s, such as polycaprolactone-based polyurethane containing 4,4′-diphenyl methane diisocyanate, which degraded to 4,4′-diamino diphenyl methane during composting [173]. All the poly(ester-urethane)s were rapidly biodegraded, but the detected toxic degradation products demonstrated that it is not enough that the materials are rapidly degraded, it is equally important to establish the degradation products and their effect on the environment.

No toxicity was detected when Bionolle/starch blends or their degradation products were tested against earthworm Eisenia fetida [174]. Cell culture testing with extracts of Bionolle and polycaprolactone showed no toxic effects on the cells [175]. Toxicity of products from polyester hydrolysis such as succinic acid, adipic acid, terephthalic acid, 1,4-butanediol, ethylene glycol, styrene glycol and 1,4-cyclohexane dimethanol was evaluated by a phytotoxicity test on germination of young radish seeds [176]. According to the phytotoxicity test the order of decreasing toxicity was styrene glycol > 1,4-cyclohexane dimethanol > adipic acid ≅ succinic acid > terephthalic acid ≅ ethylene glycol > 1,4-butanediol. In a modified Sturm test the aliphatic compounds were mineralized more easily than the aromatic compounds, especially 1,4-cyclohexane dimethanol showed exceptionally slow degradation. 1,4-Cyclohexane dimethanol also had a detrimental effect on young radish. No accumulation of aromatic oligomers was observed during composting of Ecoflex copolyester [149]. Ecotoxicological tests were also performed with a synthetic degradation medium containing high concentrations of degradation intermediates. The results showed no acute ecotoxicological effects due to the aromatic intermediates. The effect of poly(L-lactide) and poly(butylene succinate) on the growth of red pepper and tomato has also been studied [177]. Seeds of red pepper and tomato were sowed and cultivated in a soil blended with powdery PLA and PBS. PBS depressed the growth of the two plants significantly even at concentrations as low as 5%, whereas PLLA up to 35% have a negligible effect or boosted the growth of the plants. pH and number of microbial cells was not affected by the two polymers.

11 Concluding Remarks

The environmental degradation of aliphatic and aliphatic–aromatic polyesters proceeds by one or several mechanisms including chemical hydrolysis, biodegradation and thermal degradation. The structure of the polyester and the degradation environment determines the degradation rate and degradation products. The low molecular weight degradation intermediates formed during the degradation process in turn determine the environmental adaptability of the material. Monitoring the fate and identity of these products is a critical step to evaluate degradation characteristics and environmental impact of the material. The degradation product pattern is influenced by, for example, the chemical structure of the building blocks and their stereochemistry, the macromolecular architecture as well as the ability of enzymes to catalyze the hydrolysis of different ester bonds. As an example different distributions of weak linkages, crosslinking and water-solubility of the degradation intermediates all greatly influence the migration rate and degradation product pattern. Hydrolysis of aliphatic polyesters produces monomers and different oligomeric species. After aging in a biotic environment also different degradation intermediates such as carboxylic acids have been identified. The ecotoxicity tests performed on aliphatic and aliphatic–aromatic polyesters did not in most cases show any negative effects. However, if copolymerization or chain extension is applied, the new building blocks have to be chosen with care in order to retain the environmental adaptability of the polymer and its breakdown products.

The interest in biodegradable and/or renewable materials is increasing rapidly. It seems inevitable that these materials in future will have an increasing role in the management of waste and litter. The development of truly environmentally degradable polymers requires better understanding of the correlations between environmental parameters, degradation rate and breakdown mechanisms. Degradable materials need to have controlled service-life and then degrade in a reproducible and predictable fashion leaving no harmful degradation products or polymer fragments behind. Existing materials sold as degradable products are a very heterogeneous group of materials including both hydro-degradable and oxy-degradable materials. Important questions include how these materials should be tested to really ensure their degradability under relevant environmental conditions. Several factors affect the degradation process and it is difficult to isolate the effect of individual parameters. One key question is, however, what degradation products are formed and released to the environment? The ultimate fate of all individual components and the identity of intermediate and final degradation products must be well characterized in order to understand the degradation process and the environmental impact of the material. Chromatographic techniques are ideal tools for studying the degradation products released from the environmentally degradable materials.

References

1. Albertsson A-C (ed) (2002) Degradable aliphatic polyesters. Adv Polym Sci 157
2. Li SM, Garreau H, Vert M (1990) J Mater Sci Mater Med 1:123
3. Li SM, Garreau H, Vert M (1990) J Mater Sci Mater Med 1:198
4. Hakkarainen M, Albertsson A-C, Karlsson S (1996) Polym Degrad Stab 52:283
5. Cam D, Hyon S-H, Ikada Y (1995) Biomaterials 16:833
6. Hyon S-H, Jamshidi K, Ikada Y (1998) Polym Int 46:196
7. Zhang X, Wyss UP, Pichora D, Goosen MFA (1994) Bioactive Comp Polym 9:80
8. Tasaka F, Ohya Y, Ouchi T (2001) Macromolecules 34:5494
9. Schwach G, Vert M (1999) Int J Biol Macromol 25:283
10. Fischer EW, Sterzel HJ, Wegner G (1973) Kolloid-ZUZ Polym 251:980
11. Chu CC (1981) J Appl Polym Sci 15:1727
12. Williams DF (1981) Engin Med 10:5

13. Reeve S, McCarthy SP, Downey MJ, Gross RA (1994) Macromolecules 27:825
14. MacDonald RT, McCarthy SP, Gross RA (1996) Macromolecules 29:7356
15. Tokiwa Y, Jarerat A (2004) Biotechnol Lett 26:771
16. Tokiwa Y, Calabia BP (2006) Appl Microbiol Biotechnol 72:244
17. Torres A, Li SM, Roussos S, Vert M (1996) J Environ Polym Degrad 4:213
18. Karjomaa S, Suortti T, Lempiäinen R, Selin J-F, Itävaara M (1998) Polym Degrad Stab 59:333
19. Torres A, Li SM, Roussos S, Vert M (1996) J Appl Polym Sci 62:2295
20. Hakkarainen M, Karlsson S, Albertsson A-C (2000) Polymer 41:2331
21. Hakkarainen M, Karlsson S, Albertsson A-C (2000) J Appl Polym Sci 76:228
22. Renstad R, Karlsson S, Sandgren Å, Albertsson A-C (1998) J Environ Polym Degrad 6:209
23. Itävaara M, Vikman M, Venelampi O (1997) Comp Sci Utilisat 5:84
24. Day M, Shaw K, Cooney D, Watts J, Harrigan B (1997) J Environ Polym Degrad 5:137
25. Itävaara M, Karjomaa S, Selin J-F (2002) Chemosphere 46:879
26. Zellner G, Neudörfer F, Diekmann H (1994) Water Res 28:1337
27. Stridsberg K, Albertsson A-C (2000) Polymer 41:7321
28. Karlsson S, Hakkarainen M, Albertsson A-C (1994) J Chromatogr A 688:251
29. Jarerat A, Tokiwa Y, Tanaka H (2003) Biotechnol Lett 25:2035
30. Jarerat A, Tokiwa Y (2003) Biotechnol Lett 25:401
31. Vidil C, Braud C, Garreau H, Vert M (1995) J Chromatogr A 711:323
32. Braud C, Devarieux R, Garreau H, Vert M (1996) J Environ Polym Degrad 4:135
33. Reed AM, Gilding DK (1981) Polymer 22:494
34. Grijpma DW, Pennings AJ (1994) Macromol Chem Phys 195:1633
35. Li SM, Garreau H, Vert M (1990) J Mater Sci Mater Med 1:131
36. Li S (1999) J Biomed Mater Res 48:342
37. Li S, Dobrzynski P, Kasperczyk J, Bero M, Braud C, Vert M (2005) Biomacromolecules 6:489
38. Li S, Tenon M, Garreau H, Braud C, Vert M (2000) Polym Degrad Stab 67:85
39. Li S, Girard A, Garreau H, Vert M (2001) Polym Degrad Stab 71:61
40. Khabbaz F, Karlsson S, Albertsson A-C (2000) J Appl Polym Sci 78:2369
41. Makino K, Ohshima H, Kondo T (1986) J Microencapsulation 3:203
42. Tsuji H, Ikada Y (1998) J Polym Sci Polym Chem 36:59
43. Tsuji H, Mizuno A, Ikada Y (2000) J Appl Polym Sci 77:1452
44. Belbella A, Vauthier C, Fessei H, Devissaguet J-P, Puisieux F (1996) Int J Pharm 129:95
45. De Jong SJ, Arias ER, Rijkers DTS, van Nostrum CF, Kettenes-van den Bosch JJ, Hennink WE (2001) Polymer 42:2795
46. Fields RD (1973) PhD Thesis, Cornell University, New York, USA
47. Tokiwa Y, Ando T, Suzuki T (1976) J Ferment Technol 54:603
48. Oda JAY, Asari H, Urakami T, Tonomura K (1995) J Ferment Bioeng 80:265
49. Murphy CA, Cameron JA, Huang SJ, Vinopal RT (1996) Appl Environ Microbiol 62:456
50. Eldsäter C, Erlandsson B, Renstad R, Albertsson A-C, Karlsson S (2000) Polymer 41:1297
51. Lefebvre F, David C, Vander Wauven C (1994) Polym Degrad Stab 45:347
52. Ohtaki A, Akakura N, Nakasaki K (1998) Polym Degrad Stab 62:279
53. Toncheva V, Van Den Bulcke A, Schacht E, Mergaert J, Swing J (1996) J Environ Polym Degrad 4:71
54. Nishida H, Tokiwa Y (1994) Chem Lett 1293

55. Doi Y, Kasuya K-I, Abe H, Koyama N, Ishiwatari S-I, Takagi K, Yoshida Y (1996) Polym Degrad Stab 51:281
56. Mochizuki H, Hirano M, Kanmuri Y, Kudo K, Tokiwa Y (1995) J Appl Polym Sci 55:289
57. Benedict CV, Cameron JA, Huang SJ (1983) J Appl Polym Sci 28:335
58. Fields RD, Rodriguez F, Finn RK (1974) J Appl Polym Sci 18:3571
59. Renstad R, Erlandsson B, Eldsäter C, Karlsson S, Albertsson A-C (1998) J Appl Polym Sci 70:61
60. Hakkarainen M, Albertsson A-C (2002) Macromol Chem Phys 203:1276
61. Pettigrew CA, Reece GA, Smith MC, King LW (1995) J Macromol Sci Pure Appl Chem A32:811
62. Lefèvre F, Tidjani A, Vander Wauven C, David C (2002) J Appl Polym Sci 83:1334
63. Bastioli C, Degli-Innocenti F, Guanella I, Romano G (1995) JMS-Pure Appl Chem A32:839
64. Malin M, Hiljanen-Vainio M, Karjalainen T, Seppälä J (1996) J Appl Polym Sci 59:1289
65. Cha Y, Pitt CG (1990) Biomaterials 11:108
66. Liu L, Li M, Garreau H, Vert M (2000) Biomacromolecules 1:350
67. Lostocco MR, Murphy CA, Cameron JA, Huang SJ (1998) Polym Degrad Stab 59:303
68. Wang LW, Ma W, Gross RA, McCarthy SP (1998) Polym Degrad Stab 59:161
69. Gan Z, Yu D, Zhong Z, Liang Q, Jing X (1999) Polymer 40:2859
70. Tsuji H, Mizuno A, Ikada Y (1998) J Appl Polym Sci 70:2259
71. Albertsson A-C, Eklund M (1995) J Appl Polym Sci 57:87
72. DeKesel C, Vander Wauven C, David C (1997) Polym Degrad Stab 55:107
73. Höglund A, Hakkarainen M, Albertsson A-C (2007) J Macromol Sci A 44:1041
74. Hoshino A, Isono Y (2002) Biodegradation 13:141
75. Abe H, Doi Y, Aoki H, Akehata T, Hori Y, Yamaguchi A (1995) Macromolecules 28:7630
76. Sanchez JG, Tsuchii A, Tokiwa Y (2000) Biotechnol Lett 22:849
77. Ponsart S, Coudane J, Saulnier B, Morgat J-L, Vert M (2001) Biomacromolecules 2:373
78. Holland SJ, Jolly AM, Yasin M, Tighe BJ (1987) Biomaterials 8:289
79. Kanesawa Y, Doi Y (1990) Makromol Chem Rapid Commun 11:679
80. Yasin M, Holland SJ, Tighe BJ (1990) Biomaterials 11:451
81. Doi Y, Kanesawa Y, Kunioka M, Saito T (1990) Macromolecules 23:26
82. Kanesawa Y, Tanashi N, Doi Y, Saito T (1994) Polym Degrad Stab 45:179
83. Mergaert J, Webb A, Anderson C, Wouters A, Swings J (1993) Appl Environ Microbiol 59:3233
84. Budwill K, Fedorak P, Page WJ (1992) Appl Environ Microbiol 58:1398
85. Reischwitz A, Stoppok E, Buchholz K (1998) Biodegradation 8:313
86. Lenz RW (1993) Adv Polym Sci 107:1
87. Hocking PJ, Marchessault RH, Timmins MR, Scherer TM, Lenz RW, Fuller RC (1994) Macromol Rapid Commun 15:447
88. Abe H, Doi Y, Kumagai Y (1994) Macromolecules 27:6012
89. Kemnitzer JE, McCarthy SP, Gross RA (1992) Macromolecules 25:5927
90. Abe H, Doi Y (1996) Macromolecules 29:8683
91. Mukai K, Yamada K, Doi Y (1993) Polym Degrad Stab 41:85
92. Hocking PJ, Marchessault RH, Timmins MR, Lenz RW, Fuller RC (1996) Macromolecules 29:2427
93. Kumagai Y, Kanesawa Y, Doi Y (1992) Makromol Chem 193:53

94. Tomasi G, Scandola M, Briese BH, Jendrossek D (1996) Macromolecules 29:507
95. Spyros A, Kimmick R, Briese BH, Jendrossek D (1997) Macromolecules 30:8218
96. Gilmore DF, Antoun S, Lenz RW, Goodwin S, Austin R, Fuller RC (1992) J Ind Microbiol 10:199
97. Yue CL, Gross RA, McCarthy SP (1996) Polym Degrad Stab 51:205
98. Doi Y, Kanesawa Y, Tanahashi N, Kumagai Y (1992) Polym Degrad Stab 36:173
99. Eldsäter C, Albertsson A-C, Karlsson S (1997) Acta Polym 48:478
100. Eldsäter C, Albertsson A-C, Karlsson S (1999) Polym Degrad Stab 64:177
101. Kasuya K, Doi Y, Yao T (1994) Polym Degrad Stab 45:379
102. Kasuya K, Inoue Y, Yamada K, Doi Y (1995) Polym Degrad Stab 48:167
103. Abe H, Matsubara I, Doi Y, Hori Y, Yamaguchi A (1994) Macromolecules 27:6018
104. Wang Y, Inagawa Y, Saito T, Kasuya K-I, Doi Y, Inoue Y (2002) Biomacromolecules 3:828
105. Wang Y, Inagawa Y, Osanai Y, Kasuya K-I, Saito T, Matsumura S, Doi Y, Inoue Y (2002) Biomacromolecules 3:894
106. Scandola M, Focarete ML, Adamus G, Sikorska W, Baranowska I, Swierczek S, Gnatowski M, Kowalczuk M, Jedlinski Z (1997) Macromolecules 30:2568
107. Focarete ML, Scandola M, Jendrossek D, Adamus G, Sikorska W, Kowalczuk M (1999) Macromolecules 32:4814
108. Scherer TM, Fuller RC, Goodwin S, Lenz RW (2000) Biomacromolecules 1:577
109. Rychter P, Biczak R, Herman B, Smylla A, Kurcok P, Adamus G, Kowalczuk M (2006) Biomacromolecules 7:3125
110. McLellan DW, Hallings PJ (1988) J Chromatogr 445:251
111. Braud C, Devarieux R, Atlan A, Ducos C, Vert M (1998) J Chromatogr B 706:73
112. Yu J, Plackett D, Chen LXL (2005) Polym Degrad Stab 89:289
113. Albertsson A-C, Ljungquist O (1986) J Macromol Sci-Chem A23:393
114. Lindström A, Albertsson A-C, Hakkarainen M (2004) Polym Degrad Stab 83:487
115. Ahn BD, Kim SH, Kim YH, Yang JS (2001) J Appl Polym Sci 82:2808
116. Cho K, Lee J, Kwon K (2001) J Appl Polym Sci 79:1025
117. Albertsson A-C, Ljungquist O (1988) Acta Polym 39:95
118. Albertsson A-C, Ljungquist O (1988) J Macromol Sci Chem A25:467
119. Albertsson A-C, Ljungquist O (1986) J Macromol Sci Chem A23:411
120. Tokiwa Y, Suzuki T (1977) Nature 270:76
121. Fields RD, Rodriguez F (1976) Microbial degradation of polyesters. In: Sharpley JM, Kaplan AM (eds) Proc 3rd Int Biodegradation Symp. Applied Science, Barking, England, pp 775–784
122. Mochizuki M, Mukai K, Yamada K, Ichise N, Murase S, Iwaya Y (1997) Macromolecules 30:7403
123. Montaudo G, Rizzarelli P (2000) Polym Degrad Stab 70:305
124. Nikolic MS, Djonlagic J (2001) Polym Degrad Stab 74:263
125. Shirahama H, Kawaguchi Y, Aludin MS, Yasuda H (2001) J Appl Polym Sci 80:340
126. Pranamuda H, Tokiwa Y, Tanaka H (1995) Appl Environ Microbiol 61:1828
127. Suyama T, Tokiwa Y, Ouichanpagdee P, Kanagawa T, Kamagata Y (1998) Appl Environ Microbiol 64:5008
128. Teeraphatpornchai T, Nakajima-Kambe T, Shigeno-Akutsu Y, Nakayama M, Nomura N, Nakahara T, Uchiyama H (2003) Biotechnol Lett 25:23
129. Tomita K, Kuroki Y, Hayashi N, Komukai Y (2000) J Biosci Bioeng 90:350
130. Kasuya K-I, Takagi K-I, Ishiwatari S-I, Yoshida Y, Doi Y (1998) Polym Degrad Stab 59:327
131. Lindström A, Albertsson A-C, Hakkarainen M (2004) J Chromatogr A 1022:171

132. Ando Y, Yoshikawa K, Yoshikawa T, Nishioka M, Ishioka R, Yakabe Y (1998) Polym Degrad Stab 61:129
133. Kitakuni E, Yoshikawa K, Nakano K, Sasuga J, Nobiki M, Naoi H, Yokota Y, Ishioka R, Yakabe Y (2001) Environ Toxicol Chem 20:941
134. Scherer TM, Fuller RC, Lenz RW, Goodwin S (1999) Polym Degrad Stab 64:267
135. Taniguchi I, Shigeyuki N, Nakamura T, El-Salmawy A, Miyamoto M, Kimura Y (2002) Macromol Biosci 2:447
136. Rizzarelli P, Impallomeni G, Montaudo M (2004) Biomacromolecules 5:433
137. Yang J, Tian W, Li Q, Yang L, Cao A (2004) Biomacromolecules 5:2258
138. Levèfre C, Mathieu C, Tidjani A, Dupret A, Vander Wauven C, DeWinter W, David C (1999) Polym Degrad Stab 64:9
139. Allen NS, Edge M, Mohammadian M, Jones K (1991) Eur Polym J 27:1373
140. Allen NS, Edge M, Mohammadian M, Jones K (1994) Polym Degrad Stab 43:229
141. Tokiwa Y, Suzuki T (1981) J Appl Polym Sci 26:441
142. Witt U, Müller R-J, Augusta J, Widdecke H, Deckwer WD (1994) Macromol Chem Phys 195:793
143. Witt U, Müller R-J, Deckwer WD (1995) J Macromol Sci Pure Appl Chem A32:851
144. Müller R-J, Witt U, Rantze E, Deckwer WD (1998) Polym Degrad Stab 59:203
145. Witt U, Müller R-J, Deckwer W-D (1995) J Environ Polym Degrad 3:215
146. Witt U, Yamamoto M, Seeliger U, Müller R-J, Warzelhan V (1999) Angew Chem Int Ed 38:1438
147. Herrera R, Franco L, Rodríguez-Galán A, Puiggalí J (2002) J Polym Sci Part A Polym Chem 40:4141
148. Kleeberg I, Hetz C, Kroppenstedt RM, Müller R-J, Deckwer WD (1998) Appl Environ Microbiol 64:1731
149. Witt U, Einig T, Yamamoto M, Kleeberg I, Deckwer W-D, Müller R-J (2001) Chemosphere 44:289
150. Gouda MK, Kleeberg I, van den Heuvel J, Müller R-J, Deckwer W-D (2002) Biotechnol Prog 18:927
151. Chiellini E, Corti A, Giovannini A, Narducci P, Paparella AM, Solaro R (1996) J Environ Polym Degrad 4:37
152. Heidary S, Gordon B III (1994) J Environ Polym Degrad 2:19
153. Sharma PP, Gordon B III (1989) ACS Polym Prepr 30:197
154. Grzebieniak K, Slusarczyk C, Wlochowicz A, Janicki J (2002) Polimery 47:528
155. Ebato H, Oya S, Kakizawa Y, Furuta H, Arai K (1994) EU Patent 618 250 A1
156. Witt U, Müller R-J, Deckwer W-D (1996) J Environ Polym Degrad 4:9
157. Marten E, Müller R-J, Deckwer W-D (2003) Polym Degrad Stab 80:485
158. Honda N, Taniguchi I, Miyamoto M, Kimura Y (2003) Macromol Biosci 3:189
159. Höglund A, Hakkarainen M, Albertsson A-C (2007) Biomacromolecules 8:2025
160. Hakkarainen M, Höglund A, Odelius K, Albertsson A-C (2007) J Am Chem Soc 129:6308
161. Tracy MA, Ward KL, Firouzabadian L, Wang Y, Dong N, Qian R, Zhang Y (1999) Biomaterials 20:1057
162. Lindström A, Hakkarainen M (2007) J Appl Polym Sci 104:2458
163. Lee S-H, Kim SH, Han Y-K, Kim YH (2001) J Polym Sci Part A Polym Chem 39:973
164. Van Nostrum CF, Veldhuis TFJ, Bos GW, Hennink WE (2004) Polymer 45:6779
165. Benedict CV, Cook WJ, Jarrett P, Cameron JA, Huang SJ, Bell JP (1983) J Appl Polym Sci 28:327
166. Pitt CG, Gratzl MM, Himmel GL, Surles J, Schindler A (1981) Biomaterials 1:215
167. Huffman KR, Casey DJ (1985) J Polym Sci Polym Chem 23:1939

168. Albertsson A-C, Löfgren A (1994) J Appl Polym Sci 52:1327
169. Li S, Vert M (1994) Macromolecules 27:3107
170. Bowmer CT, Hooftman RN, Hanstveit AO, Venderbosch PWM, van der Hoeven N (1998) Chemosphere 37:1317
171. Alauzet N, Garreau H, Bouché M, Vert M (2002) J Polym Environ 10:53
172. Tuominen J, Kylmä J, Kapanen A, Venelampi O, Itävaara M, Seppälä J (2002) Biomacromolecules 3:445
173. Tosin M, Degli-Innocenti F, Bastioli C (1998) J Environ Polym Degrad 6:79
174. Jayasekara R, Sheridan S, Lourbakos E, Beh H, Christie GBY, Jenkins M, Halley PB, Mcglashan S, Lonergan GT (2002) Int Biodet Biodegrad 51:77
175. Dang M-H, Birchler F, Wintermantel E (1997) J Environ Polym Degrad 5:49
176. Mal-Nam K, Boo-Young L, Ick-Mo L, Han-Sup L, Jin-San Y (2001) J Environ Sci Health A Toxic/Hazard Subst Environ Eng A36:447
177. Kim M-N, Shin J-H, Im S-S (2003) J Polym Environ 11:101

Adv Polym Sci (2008) 211: 117–157
DOI 10.1007/12_2007_127

Published online: 29 January 2008

Chromatographic Analysis of Antioxidants in Polymeric Materials and Their Migration from Plastics into Solution

Clara Strandberg · Ann-Christine Albertsson (✉)

Department of Fibre and Polymer Technology, Royal Institute of Technology, Teknikringen 56-58, 100 44 Stockholm, Sweden
aila@polymer.kth.se

Abstract The accurate and reliable measurement of antioxidant content in polymers by chromatographic techniques, e.g., liquid chromatography (LC) is an important tool in quality and manufacturing control, troubleshooting, and material or vendor identification. This chapter describes characterization by chromatographic techniques and sample preparation methods to identify and quantify antioxidants and their degradation products in polymeric materials. The major difficulty in the characterization is usually not the analytical method but rather the separation of the antioxidants from the polymer matrix. The separation can be performed by various extraction methods. Traditionally, Soxhlet or polymer dissolution methods are used. These methods have several disadvantages. Therefore, more complex and efficient methods with the possibility of working at elevated temperatures and pressures have been developed, i.e., microwave-assisted extraction (MAE), supercritical fluid extraction (SFE) and accelerated solvent extraction (ASE), which are discussed here. These methods make it possible to study the relation between antioxidant efficiency and the release of degradation products from polymers. Another interesting aspect is the study of migration of antioxidants from plastic packaging into the surrounding media.

Keywords Antioxidants · Gas chromatography (GC) · Liquid chromatography (LC) · Microwave assisted extraction (MAE) · Polymers

Abbreviations

ACN	Acetonitrile
α-Tocopherol	3,4-Dihydro-2,5,7,8-tetramethyl-2-(4,8,12-trimethyltridecyl)-2*H*-1-benzopyran-6-ol
ASE	Accelerated solvent extraction
ATR	Attenuated total reflection
BHA	*tert*-Butyl-hydroxyanisole
BHT	2,6-Di-*tert*-butyl-*p*-cresol (or butylated hydroxyl toluene)
DBP	2,4-Di-*tert*-butylphenol
DCM	Dichloromethane
DSC	Differential scanning calorimetry
EAA	Poly(ethylene-*co*-acrylic acid)
ESE	Enhanced fluid extraction
FTIR	Fourier transform infrared spectroscopy
GC	Gas chromatography
GC-MS	Gas chromatography–mass spectroscopy
HDPE	High-density polyethylene
^{1}H-NMR	Hydrogen nuclear magnetic resonance spectroscopy
Hostanox O3	Bis-(3,3-bis-(4′-hydroxy-3′-*tert*-butylphenyl) butanoic acid)-glycol ester
Hostanox O10	Pentaerythryl-tetrakis 3-(3,5-di-*tert*-butyl-4-hydroxy-phenyl)propionate
HPLC	High-performance liquid chromatography
HPLC-ESI-MS	high-performance liquid chromatography electrospray ionization mass spectrometry
HS-SPME	Headspace solid-phase microextraction
IPA	Isopropyl alcohol
Irganox 1076	Octadecyl 3,5-di-*tert*-butyl-4-hydroxyhydrocinnamate
Irganox 1010	Pentaerythritol tetrakis(3-(3,5-di-*tert*-butyl-4-hydroxyphenyl)propionate)
Irgafos 168	Tris(2,4-di-*tert*–butylphenyl)phosphate
Irganox 1081	6-6′-Di-*tert*-butyl-2-2′-tiodi-*p*-kresol
LC	Liquid chromatography
LLE	Liquid–liquid extraction
LLDPE	Linear low-density polyethylene
LSC	Liquid scintillation counting
MS	Mass spectroscopy

MAE	Microwave assisted extraction
M_w	Molecular weight
OIT	Oxygen induction time
PBD	Polybutadiene
PBT	Poly(1,4-butylene terephthalate)
PE	Polyethylene
PE-core/toc	PE stabilized with 0.1 wt-% α-tocopherol and 0.1 wt-% core-shell particles
PE-EAA/toc	PE stabilized with 0.1 wt-% α-tocopherol and 0.1 wt-% EAA
PE-Irg1076	PE stabilized with 0.1 wt-% Irganox 1076
PE-toc	PE stabilized with 0.1 wt-% α-tocopherol
PE-ref	Un-stabilized PE
PET	Polyethylene terephthalate
PP	Polypropylene
PS	Polystyrene
PVC	Poly(vinyl chloride)
RSD	Relative standard deviation
SEC	Size exclusion chromatography
SFE	Supercritical fluid extraction
SFC-FID	Supercritical fluid chromatography flame ionization detection
SLE	Solid–liquid extraction
SPE	Solid-phase extraction
SPME	Solid-phase microextraction
THF	Tetrahydrofuran
TLC	Thin-layer chromatography
UAE	Ultrasonic-assisted extraction
Ultranox 626	Bis(2,4-di-*tert*-butylphenyl)pentaerytritoldiphosphite
UV	Ultraviolet radiation

1 Introduction

The important factors that govern polymer production are cost and end-use performance, which in many cases depends on the efficiency of antioxidants. This chapter gives an overview of the use of liquid and gas chromatography together with various extraction methods for the characterization of antioxidants and their degradation products in polymers, and of the relation between antioxidant efficiency, polymer stability and release of degradation products. When plastic materials are used as food packaging, antioxidants or their degradation products can migrate from the plastics to foodstuffs during storage. From a chemical point of view, the antioxidants belong to a various classes of compounds. Some of them are known to be potential hazardous compounds [1] e.g., allergic reactions (contact dermatitis) to plastic gloves have been seen to be related to antioxidants [2]. Therefore, identification and quantification quality control and a greater knowledge of presumed harmful antioxidants substances in plastics is necessary. Liquid chromatography (LC), gas chromatography (GC) or UV spectroscopy is usually used for identification and quantification of the antioxidants. High-performance LC (HPLC) is often used for accurate identification of low-levels of antioxidants. GC techniques, including GC-mass spectrometry (GC-MS) have proven to be essential in several applications e.g., in predictions of material performance [3], in the identification of polymer degradation [4, 5], in finding degradation mechanisms of polymers [4] and in purity control of products [6, 7]. The analyses of antioxidants in polymers are associated with particular problems that are due mainly to the high reactivity and the low stability of most antioxidants, the character of the polymer matrix and the low concentration of antioxidants. Direct analysis of the antioxidants in the polymer matrixes is difficult, not only because a high-molecular weight polymer cannot be analyzed and injected into either a liquid or a gas chromatograph, but also since many polymers have a low solubility in common solvents. As a result, the antioxidants initially have to be separated from the polymer matrix by an extraction step. Different polymers and unique combinations of antioxidants require that each analysis has to be specifically designed. There is currently a particular progress in the development of new more advanced sample preparation methods. The advance in sample preparation is driven by both scientific considerations, i.e. extraction without transforming the additive, and by practical considerations such as time and cost efficiency.

2 The Function and Nature of Antioxidants

2.1 Stabilization of Polymers

Antioxidants are used in polymeric materials both for processing and for long-term stabilization to prevent degradation of the material. They are usually efficient even at low concentrations. Degradation (i.e. ageing of a material) leads to discoloration and deterioration in surface properties. More serious is a loss in mechanical properties such as tensile strength, elongation, etc. and, consequently, a decrease in the lifetime and usefulness of the polymer. Degradation is initiated during the processing and manufacturing, e.g. in the extrusion of a polymer due to high temperature (Δ) and shear (θ) (Scheme 1) [8]. In a stabilized material, consumption of the antioxidant occurs instead of degradation and oxidation of the polymer.

Different types of antioxidants are used for the stabilization of polymers: *H*-donors, radical scavengers and hydroperoxide decomposers. They interfere in different ways during the auto-oxidation cycle of polymeric materials (Scheme 2).

H-donors, e.g. phenolic antioxidants, hinder the reaction between alkoxy radicals and the polymer backbone. Radical scavengers react mainly with

$$\text{a)}\quad R\text{-}H \xrightarrow{\Delta,\theta} R^{*}$$

$$\text{b)}\quad R^{*} + O_2 \rightarrow ROO^{*}$$

$$\text{c)}\quad ROO^{*} + RH \rightarrow ROOH + R^{*}$$

Scheme 1 *a* Initiation of degradation of a polymer chain (R-H) by heat or light with the formation of a carbon-centered radical (R*). *b* Oxidation of the polymer chain by formation of a peroxyl radical (ROO*). *c* Formation of a hydroperoxide (ROOH) and R*, leading to auto-oxidation

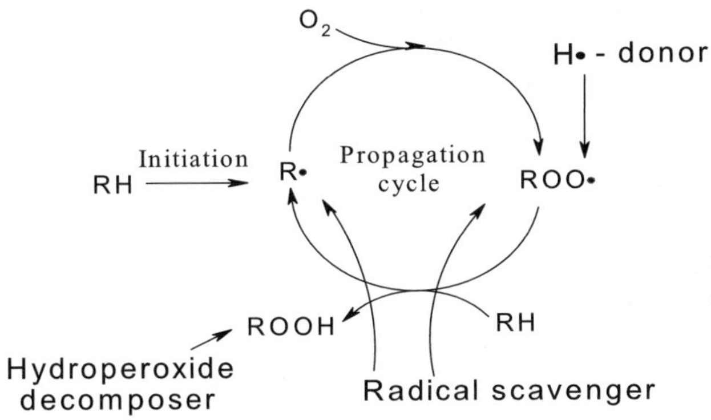

Scheme 2 Inhibition of auto-oxidation by different antioxidant mechanisms

alkyl radicals and with alkoxy radicals, whereas hydroperoxide decomposers transform hydroperoxides into non-radical products, suppressing the formation of alkoxy and hydroxyl radicals.

2.1.1 Phenolic Antioxidants

Sterically hindered phenols (Scheme 3) are the most widely used antioxidants in polymers. They are characterized by having a labile hydrogen atom with low dissociation energy and by not reacting further by hydrogen abstraction from the polymer backbone. They react primarily with alkoxy radicals (Scheme 4) and their oxidation products, quinones, react with peroxy radicals.

2.2 Demands on Antioxidants

An antioxidant should be chemically able to interfere with the oxidation reactions, and should resist its own degradation and loss by migration, leaching or precipitation on the surface. The rate of loss of an antioxidant is determined by its volatility, solubility, and diffusion rate, i.e. by its mobility in

Irganox 1010

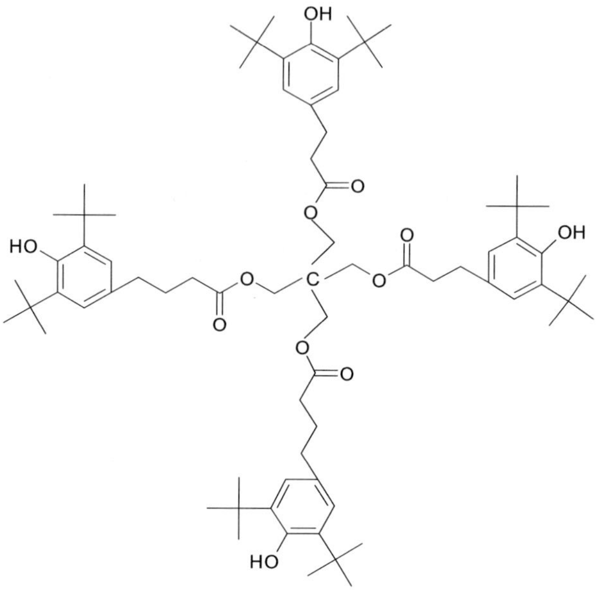

Irganox 1076

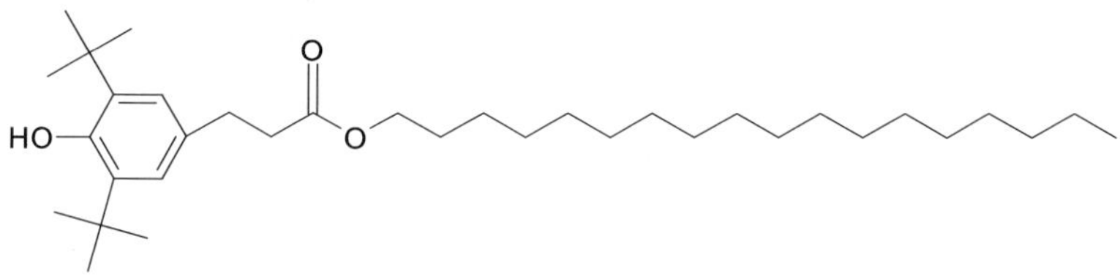

Scheme 3 Example of two commonly used phenolic antioxidants, Irganox 1010 and Irganox 1076

$$ROO^* + Ar\text{-}OH \rightarrow ROOH + Ar\text{-}O^*$$

$$R^* + Ar\text{-}O^* \rightarrow \text{Inert product}$$

Scheme 4 Chain-breaking mechanisms involving the H-donating phenolic antioxidant (Ar – OH), peroxyl radical (ROO*), hydroperoxide (ROOH), phenolic radical (Ar – O*) and carbon-centered radical (R*)

the polymer. The efficiency of an antioxidant is dependent on its consumption, diffusion and loss from the surface, where most of the oxidation occurs. The rate of diffusion of an antioxidant is dependent on its specific volume, on the total free volume of the system and on the mobility of the polymer chain-segments [9]. The solubility of an antioxidant in a polymer depends on its specific volume, physical state, possible self-association of the antioxidant, polymer–antioxidant interaction [9], the temperature and the pressure [10]. Most synthetic antioxidants are polymorphous substances forming various physical structures below their melting point [11]. Their solubility depends strongly on the thermal history of the polymer-additive system and on the surrounding medium. A relationship has been seen between the thermo-oxidative stability of a polymer and the separation behavior of antioxidants, and this separation has been shown to be directly correlated with the solubility, depending on the composition of the antioxidant and of the polymer blend [12].

3
Determination of Antioxidant Content in Polymers

3.1
General Considerations

The analysis of antioxidants is based on their non-volatile nature, their modest molecular weight and the fact that they absorb light due to their chromophore groups. In works concerning the separation of additives from polyolefins the main difficulty is generally to achieve the extraction and recovery in a reasonable time. Differences in shape, size or thickness of a polymer film or the presence of other additives may significantly affect the result of the extraction and analysis of the antioxidant content in a polymer. Consequently, method development has to be performed to find standardized methods.

3.2
Analytical Techniques

3.2.1
Direct Methods

Direct methods of analysis such as ultraviolet (UV) absorption, infrared spectroscopy (IR), fluorescence, phosphorescence [13], X-ray fluorescence [14–16] and thermal analysis [17] have been reported. However, these methods generally lack specificity [18]. In Fourier transform IR (FTIR), overlapping bands of other species may interfere with the absorbance bands of the analyte, and in UV analysis the absorbance bands of different antioxidants can be very similar. UV and FTIR analysis are especially useful techniques when an antioxidant system is already known. X-ray fluorescence and elemental analysis are fast and useful techniques for the determination of antioxidants containing phosphorus or sulfur. The measurement of oxygen consumption during a cumene-initiated oxidation of sterically hindered amine stabilizers has been used to determine the relative content of antioxidants in polypropylene (PP) [19]. Nuclear magnetic resonance spectroscopy (NMR) analysis has also been used for analysis of additives and their degradation products in polymers [20]. With both FT-IR and NMR, quantitative results are directly obtained.

3.2.2 Chromatographic Techniques

The amount and type of stabilizer determine the service life and performance of many polymers. A complete characterization includes both the identity of the compounds and their levels in a product. Chromatographic techniques offer accurate and reliable identification of antioxidant content in polymers and are therefore an important tool for quality and manufacturing control, troubleshooting, and material or vendor detection. In chromatography, the components in a sample are separated by distribution between two phases, of which one is stationary and the other is moving, i.e. mobile. The distribution of analytes between the two phases depends on their relative affinities for the phases, as determined by molecular structures and intermolecular forces. The choice of chromatographic technique depends on the nature of the analyte, on the sample matrix and on the purpose of the analysis. A quantitative analysis of the amount of an analyte is achieved by comparison with suitable standards, assuming a linear relationship between the peak areas of the standards and their concentrations. Antioxidants are separated from a mixture and subsequently identified and quantified by liquid, gas or supercritical fluid chromatography. Chromatographic techniques, particularly thin-layer chromatography (TLC), were used at an early stage for the identification of antioxidants in rubber [21]. TLC was also used to determine antioxidants in polyethylene (PE) [22]. TLC is the least expensive chromatographic technique and it is still a valuable tool for screening the antioxidant content in a polymer [23]. Size exclusion chromatography (SEC) is another technique employed for the analysis of polymer additive systems [24–27]. The analysis time in SEC is usually longer than that required for LC analysis. In SEC, molecules are separated based on their molecular weight (M_W), whereas in LC the separation of compounds is governed by both their polarity and size.

3.2.2.1 Liquid Chromatography

In liquid chromatography (LC), the separation of analytes is based on their distributions between a mobile liquid phase and a stationary phase. The efficiency of the separation depends on the chemical properties of the analytes, on the structure and pore size of the stationary phase, on the length and the inner diameter of the column, and on the mobile phase structure. LC is an accurate and reproducible technique for the identification and quantification of antioxidants [28–35]. Analytes with a high boiling point and thermal sensitive analytes are especially suited for LC analysis, since the analyte is not transformed during the analysis. Some additives lack UV absorbing or fluorescent chromophores and MS analysis is then frequently used subsequent to the LC analysis. MS has the advantage of being specific and of having excellent limits of detection. It also eliminates purification and clean-up steps. Gradient LC is a practical method for the separation and characterization of complex antioxidant packagings. When phosphatic antioxidants are analyzed, care must be taken since they are intermittently unstable in water. Normal-phase HPLC analysis using a heptane : methyl chloride gradient has been shown to have a better performance than both reverse-phase HPLC (with ACN : THF, 75 : 25, as mobile phase) and SEC (with methyl chloride as mobile phase) [36]. A separation and characterization method based on a coupling between LC and off-line FTIR has been shown to give a limit of detection of the phosphatic antioxidant Irgafos P-EPQ of about 40 ng and a relative standard deviation (RSD) of 4.4% [37]. Irganox 1076, Cyasorb 531, Tinuvin 327 and Tinuvin 328 from PP and polyvinyl chloride (PVC) have also been analyzed by semi-online coupled LC and FTIR [38]. A spray-jet interface was used to deposit the effluent from a narrow-bore LC column on a zinc selenide window. The limits of detection were in the nanogram range. Online coupling of SEC and normal-phase HPLC has been used for the analysis of BHT, Irganox 1076, Tinuvin 326, Tinuvin 327, Cyasorb UV 9 and Cyasorb UV 1084 from polystyrene (PS) [39]. An automatic three-way switch valve was placed between the two columns. The antioxidants and UV stabilizers were separated from the polymer in the SEC column, and then identified and quantified in the HPLC column. Detection limits were about 0.1 mg/mL for BHT, Tinuvin 326 and Tinuvin 327, 0.2 mg/mL for Irganox 1076, and 1.1 mg/mL for Cyasorb UV 9 and Cyasorb UV 1084. The RSD values of the method were below 4%. Crozier et al. [40] studied the effectiveness of five reverse-phase HPLC columns for the analysis of flavonoids. The best results were obtained with a C18 Nova-Pak, a C18 Symmetry and a C18 Genesis column.

3.2.2.2
Gas Chromatography

In gas chromatography (GC), the separation of analytes is based on a distribution between a mobile gas phase and a stationary phase. It is a technique with high resolution, high selectivity and high sensitivity. A range of different detectors can be used. This technique has nevertheless certain limitations in the characterization of antioxidants, since many antioxidants are thermally sensitive. GC analysis is also limited to lower molar mass compounds, additives, oligomers, etc., since the analytes must have a boiling point below the maximum analysis temperature. Higher molecular mass compounds have low volatility and some antioxidants, e.g. α-tocopherol, even have no boiling point.

Using GC-mass spectrometry (GC-MS) and HPLC, Irganox 1010 and some other alkylated phenols and their degradation products have been identified in fumes from hot melt glue [41]. Using tandem MS and LC, the antioxidant content of bis-thiophenols in a cross-linked elastomer polymer was determined after extraction of the antioxidants from the polymer by THF at room temperature for 16 h [2]. The tandem MS made it possible to distinguish between three isomeric structures of one of the antioxidants. Both the MS and the HPLC results showed that the concentration of the antioxidant in the polymer was 1 wt-%. GC-MS together with thermal desorption has been used to identify volatile compounds, e.g. BHT from therapeutic tubing made of PVC [42].

3.2.2.3 Supercritical Fluid Chromatography

In supercritical fluid chromatography (SFC), a supercritical fluid, usually carbon dioxide, is used. An SFC column resembles a GC column or a packed LC column. The operating temperature is lower than that in GC, which makes it more suitable for the analysis of thermally labile compounds. Another advantage is that water is not present and this is particularly valuable for the analysis of phosphatic antioxidants. In addition, high molecular mass antioxidants can be analyzed with SFC. SFC and FTIR have been used for the analysis of light stabilizers (UV absorbers) and antioxidants [43]. The eluted compounds were deposited on a KBr window after they had passed through a capillary restrictor. The limit of detection was around 100 ng.

3.2.3 Comparison Between Chromatography and Thermal Analysis

As indicated in the previous sections, the antioxidant content in plastic material is often determined by chromatographic methods. Another widely used technique for polymer characterization is thermal analysis with differential scanning calorimetry (DSC). When the oxygen induction time (OIT) for a sample containing a phenolic antioxidant is measured, a significant oxidative exothermic response is obtained in the DSC when all the phenolic antioxidant in a sample is consumed. The OIT is thus directly related to the antioxidant content in the material and to the stabilizing function, i.e. the antioxidant efficiency in the sample, if the consumption of phenolic antioxidants obeys zero-order kinetics at the temperature used [44]. Table 1 shows the amount of the antioxidant Irganox 1081 in polyethylene (PE) determined by HPLC and extraction by microwave assisted extraction (MAE),

Table 1 Residual antioxidant amount of Irganox 1081 determined by OIT and by HPLC/MAE (Strandberg and Albertsson, unpublished data)

Time (days)	OIT (min) Water	Nitrogen	HPLC/MAE (%) Water	Nitrogen
0	35.7	35.7	80.2	80.2
83	42.2	35.5	66.6	69.1
218	39.9	31.3	55.3	67.5

compared to the amount derived from OIT values, during aging in water or in nitrogen (Strandberg and Albertsson, unpublished data).

HPLC analysis of the extracts from the samples aged in water showed that about 30% of Irganox 1081 was lost after 7.3 months (218 days), whereas the OIT value shows that the stabilizing effect of the antioxidant in the material was more or less the same as for the un-aged sample. The samples aged in the nitrogen environment show a greater agreement between the chromatographic and the OIT results. The same trends are visible in Fig. 1. The two different methods of measuring the antioxidant content did not agree very well. An explanation of this is probably that degradation products of Irganox 1081 acted like antioxidants and gave a contribution to the OIT. In HPLC, they have a different retention time from Irganox 1081 and were therefore not included in the quantification. Nevertheless, the values agreed better

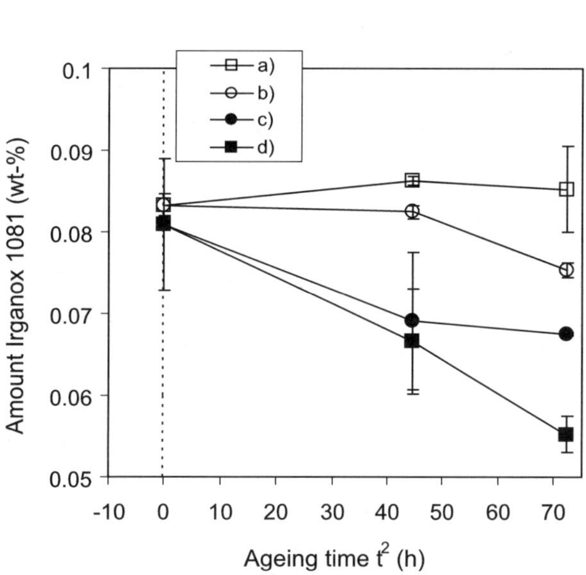

Fig. 1 Residual antioxidant amount (%) of Irganox 1081 determined by OIT during ageing in **a** H_2O (□), **b** N_2 (○), by HPLC during ageing in **c** N_2 (•), and in **d** H_2O (■)

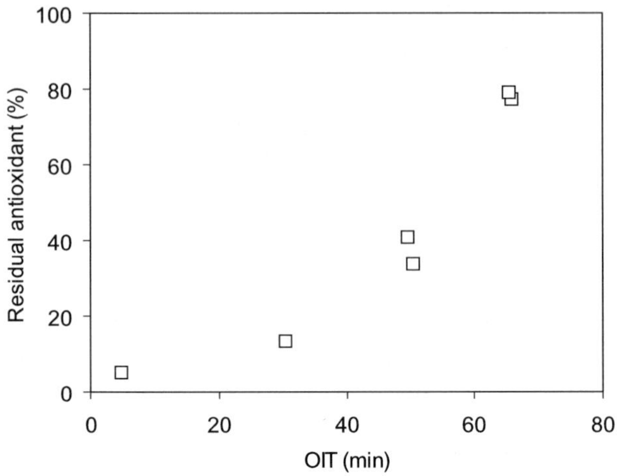

Fig. 2 Residual antioxidant content (%) of Santonox R in PE determined by HPLC (extraction by MAE), plotted versus the corresponding OIT values, for samples aged in oxygen-free water at 90 °C. © owned by the first author [46]

for the PE aged in nitrogen than for the PE aged in water. This is probably due to less formation of degradation products in this environment.

Similar results are shown in Fig. 2 for the antioxidant Santonox R aged in oxygen-free water at 90 °C [46].

3.3 Extraction Techniques

Both a qualitative and a quantitative analysis of antioxidants in polymers demand a procedure for sample preparation. Different methods are available for the extraction of antioxidants and other additives from polymers. When selecting an extraction technique, the efficiency of the extraction, the stability of the antioxidants under the extraction parameters, the time required for the extraction, the cost of the equipment, and the solvent used have to be considered. An extraction should isolate an antioxidant without transforming it, i.e. degrading it. Diffusion of an antioxidant from a polymer is dependent on several parameters, e.g. sample thickness or particle size [47], temperature, pressure, solvent, type of antioxidant and type of polymer. Larger antioxidants with higher molecular mass and large molecular size are more difficult to extract from polymers than smaller ones. The diffusion coefficient of additives in polymers at 40 °C is around 10^{-10} $cm^2\,s^{-1}$, which means that for an extraction of approximately 20 min a sample thickness of 0.3 mm is suitable. Extraction from a thin film or ground particles is therefore necessary for more efficient extraction. The grinding is done with the frozen polymer to diminish loss of antioxidant.

3.3.1
Traditional Extraction Techniques

Soxhlet extraction (developed by F. Soxhlet in 1879) and polymer dissolution are both examples of traditionally used solid–liquid extraction techniques.

3.3.1.1
Soxhlet Extraction

In Soxhlet extraction, low molecular weight compounds such as antioxidants are concentrated from polymeric materials. The extraction is carried out at a high temperature in a Soxhlet laboratory glass with a special cellulose container: a Soxhlet sock that allows continuous reflux of solvent through the sample. The high temperature may cause degradation of sensitive antioxidants. A large sample mass can be extracted and the extraction is said to be independent of the sample matrix [48]. If the extracts are too dilute, further concentration, e.g. by evaporation, may be needed. This is time-consuming and may lead to loss of volatile compounds. A pre-column is usually necessary if HPLC analysis is performed after the extraction, due to remains of low molecular weight polymer (oligomers) in the extract, but no filtration of the sample is needed before the analysis.

As in many other extraction techniques, the solvent is often a good swelling agent for the polymer matrix. The solvent has to be chosen with care. Chloroform is frequently used, but it does not always have the highest efficiency. Wims and Swarin [27] have shown that in the extraction of antioxidants from talc-filled PP with chloroform a time of 72 h was required, whereas with THF a shorter time of 24 h was sufficient. The extracts were analyzed by SEC or by HPLC. A drawback of the extraction method is that large solvent volumes are required, which make it environmentally hazardous and expensive because of purchasing and disposal costs. The extraction is rather time-consuming since the extraction process is slow; at least 2 h are usually needed, depending on the size of the antioxidant concerned. In methylene chloride [49, 50] or in methanol [51], Soxhlet extraction of antioxidants has been shown to need between 5 and 48 h. Normally, the extraction time is between 6 and 48 h [28, 50, 52].

Lichtentaler and Ranfeldt have shown that the size of the antioxidant has an impact on the extraction efficiency. They Soxhlet-extracted five phenolic antioxidants (BHT, Irganox 1010, Irganox 1076, Santonox R, Ionox 330) from a PE powder with chloroform for a time between 2 and 5 h, and analyzed the extracts by both isocratic HPLC analysis in *n*-hexane and by gradient HPLC in *n*-hexane : DCM (0–30%) [53]. In the case of BHT, only 2 h of extraction were needed to achieve an extraction yield of 95%, whereas the larger antioxidant Irganox 1076 required a time of 3.5 h for 90% recovery, and Irganox 1010 required a time of 5 h for a yield of about 85% (Fig. 3).

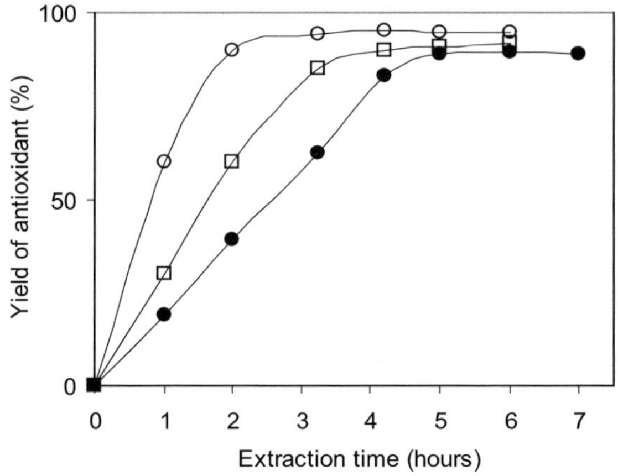

Fig. 3 Yields (%) of the antioxidants Irganox 1010 (•), Irganox 1076 (□) and BHT (○) as a function of Soxhlet extraction time. Analysis by HPLC: BHT in *n*-hexane, Irganox 1076 in *n*-hexane : DCM (8 : 2) and Irganox 1010 in *n*-hexane : DCM (4 : 6). Redrawn from [53]. With permission from Elsevier July 2007

In order to degrade the antioxidants, the PE powder was subjected to thermal aging at 200 °C for a time of 15–90 min, and the samples were then exposed to UV radiation (sunlight) for 14 days. More than 20 degradation products of BHT, most of them from photo-oxidation, were determined by gradient elution. The main ones were 2,6-di-*tert*-butyl-*p*-quinomethane and 2,6-di-*tert*-butyl-4-hydroperoxy-2,5-cyclo-hexadiene-1-one (Scheme 5). In the chromatograms for the aged and extracted PE samples containing Irganox 1076 and Irganox 1010, several additional peaks, probably originating from degradation products of the antioxidants, could also be seen.

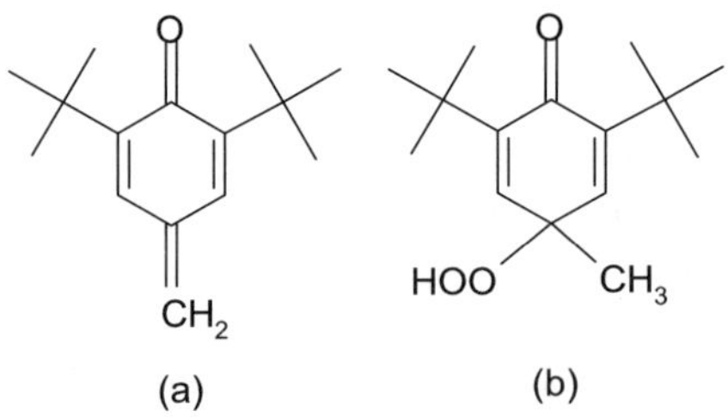

Scheme 5 Degradation products of BHT: *a* 2,6-di-*tert*-butyl-*p*-quinomethane, *b* 2,6-di-*tert*-butyl-4-hydroperoxy-2,5-cyclo-hexadiene-1-one

The thickness of the samples is also crucial. It has been observed that a 40% lower extraction yield of certain antioxidants is obtained from PE granulates than from a thin film of the same material [54]. In Soxhlet extraction of un-ground PVC, a yield of 59% of Tinuvin 320 was detected [55], whereas from the ground polymer a 97% yield was obtained [47].

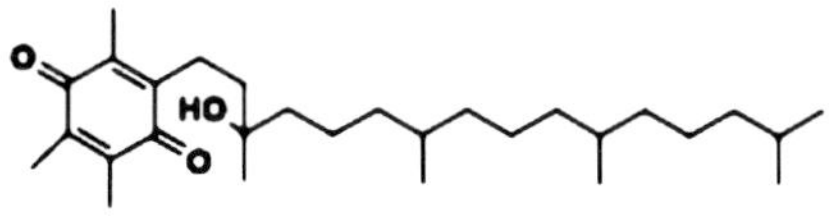

(I)

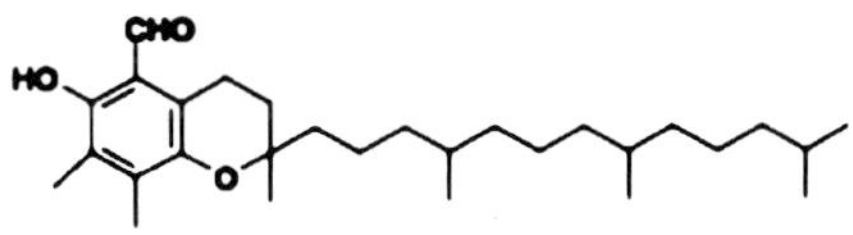

(II)

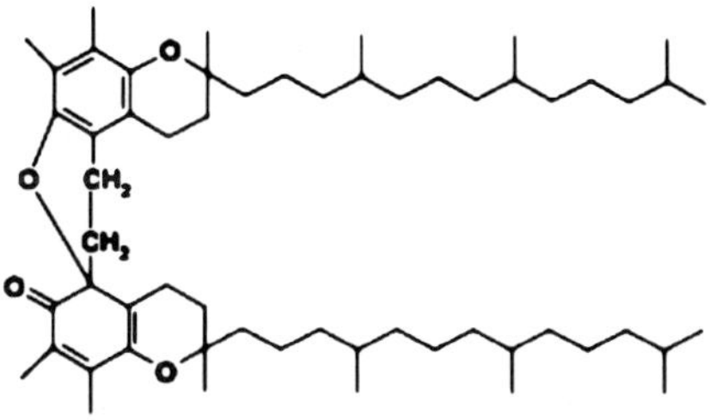

(III)

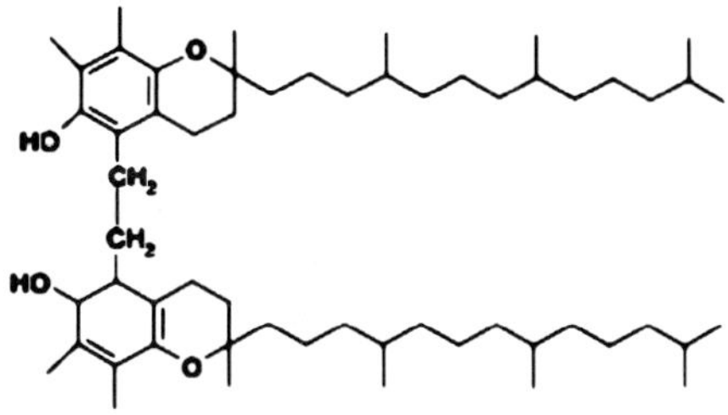

(IV)

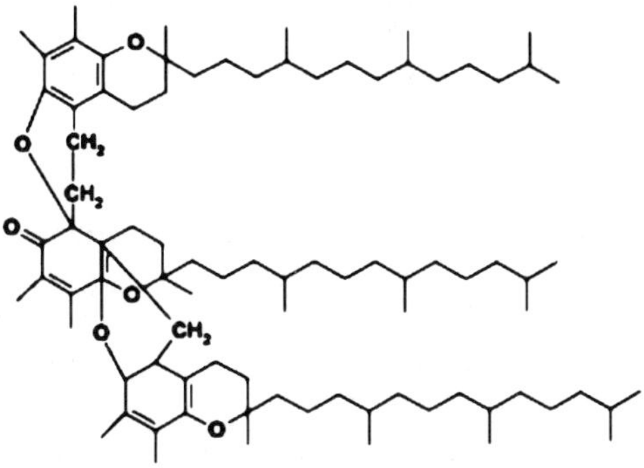

(V)

Scheme 6 Oxidation products from α-tocopherol: quinonoids (*I*), aldehydes (*II*), dimers (*III*, *IV*) and trimers (*V*)

Al-Malaika et al. [56–60] have extracted degradation products of α-tocopherol, formed during the melt processing of PP, by Soxhlet extraction for a time of 6 h with DCM. The concentrations of the different degradation products were calculated from the extinction coefficients in HPLC analysis in *n*-hexane [56]. The HPLC analysis in *n*-hexane was used to fractionate the extracts; the fractions were then analyzed by UV, FTIR, ^{1}H-NMR and MS. Several oxidation products (Scheme 6) were found: quinonoids (I), trimers (V), dimers (III, IV) and aldehydes (II).

3.3.1.2
Dissolution of Polymer

Traditionally, polymer dissolution by refluxing of a polymer in toluene, followed by precipitation of the polymer by another solvent, e.g. ethanol, has been used for the extraction of antioxidants [61], although the method is

Table 2 Extraction yields, RSD and limit of detection of BHT, Irganox 1076 and Irganox 1010 after extraction. Permission for reproduction from ACS [36] (of re-analyzed data)

Antioxidant	Yield (%)	RSD (%)	Limit of detection (wt-%)
BHT	100	3.7	0.001
Irganox 1076	107	1.6	0.004
Irganox 1010	96	11.7	0.007

rather time-consuming [15]. Schabron and Fenska [36] extracted the antioxidants BHT, Irganox 1076 and Irganox 1010 from PE granulates using naphthalene (50 mL) at 110 °C for 30 min. The amount of antioxidant was determined by HPLC after filtration of the extracts. No internal standard was used. The extraction yields, RSD and limits of detection of the antioxidants are shown in Table 2. One problem encountered was evaporation of the solvent during the extraction, which probably meant that the yield determined was higher than the true yield. This method was also used for the extraction of Santonox R, Ethyl 330, Goodrite 3114, and Topanol CA from PE and PP [62]. Additives from PP have also been extraction by polymer dissolution with naphthalene at elevated temperatures [18].

Several antioxidants has been extracted by refluxing in acetone from different types of PE for at least 2 h and analyzed by GC analysis [51]. The limits of detection were 50 ppm with RSD values of 2%. Two low boiling solvents, carbon tetrachloride and THF, were chosen for the extraction of BHT, Irganox 1076, Tinuvin 327 and of Cyasorb 531 from PE and PP [63]. The materials had initially been processed by solution blending or extrusion. The extractions were performed under reflux for 2 h and analyzed by HPLC. The method gave an antioxidant yield of almost 100%. It was seen that antioxidants were lost during the processing by extrusion. The effects of different processing methods on polymer properties and on the residual antioxidant content have been shown by Strandberg and Albertsson [17].

Degradation products of Irganox 1330 have been extracted from irradiated PP samples by refluxing with chloroform for at least 2 h followed by particle-beam LC-MS analysis. The compounds with quinone methide structures, e.g. 1,3,5-trimethyl-2,4,6-tris(3′,5′,di-*tert*-butyl-4-hydroxybenzyl)benzene were quantified [64].

3.3.2
Ultrasonic Assisted Extraction

The rate of transfer of antioxidants from a solid polymer into a solvent is increased by ultrasound waves. Ultrasonic-assisted extraction (UAE) is a rather fast and efficient method [65]. Usually, it takes less than 1 h when the sample is stirred. Solvent mixtures like cyclohexane : IPA or DCM : cyclohexane have been used [30, 31], but also DCM or chloroform alone [49, 66], with extraction times ranging from 15 min to 1 h. Haider and Karlsson [66] used UAE to extract a polymeric photo-antioxidant (Chimassorb 994), Irganox 1010, and Irgafos 168 from PE by chloroform at a temperature of 60 °C, although chloroform has a boiling point of 61 °C. The extraction was performed in closed vessels in an ultrasonic bath for a time of 5–60 min and the extracts were analyzed by UV spectra measurement after filtration or by HPLC. An extraction yield of 100% of Irgafos 168 was obtained after less than 15 min, whereas the higher molecular mass antioxidants needed more time. The full recovery of Irganox 1010 required 45 min and Chimassorb 994 1 h.

3.3.3
Techniques Based on Higher Pressure

An increasing demand for new extraction techniques, susceptible to automation, with shorter extraction times and reduced solvent consumption, giving reduced sample preparation costs and preventing pollution in laboratories, has been seen in recent decades. This progress in sample preparation has resulted in new techniques such as microwave assisted extraction (MAE), supercritical fluid extraction (SFE) and accelerated solvent extraction (ASE). These techniques are similar in that they involve working at elevated temperatures and pressures, which improves the speed of the extraction procedure.

3.3.3.1
Microwave Assisted Extraction

In microwave assisted extraction (MAE), microwave energy accelerates the partition, i.e. the mass transfer of an analyte from a sample matrix into a solvent, by directly heating the solution. The extraction is performed at an elevated temperature in a closed vessel. The major benefits are the shorter extraction time, reduced consumption of organic solvents and increased sample throughput. However, there is a need for an additional filtration step and, if the extract is dilute, further concentration, e.g. by evaporation or SPE, may be needed before analysis.

Freitag and John [67] and Nielson [31] extracted antioxidants from PP and PE using domestic microwave ovens. Over 90% of the substances were recovered from the powdered polymers within 6 min, using an acetone : heptane (1 : 1) mixture [67]. A mixture of cyclohexane : IPA (1 : 1) as extracting solvent with an extraction time of 20 min and stirring every 5 min also gave good results [31]. When the resins were extracted as pellets, full recoveries were achieved using MAE, except for Irganox 1010 (only 50% recovery without

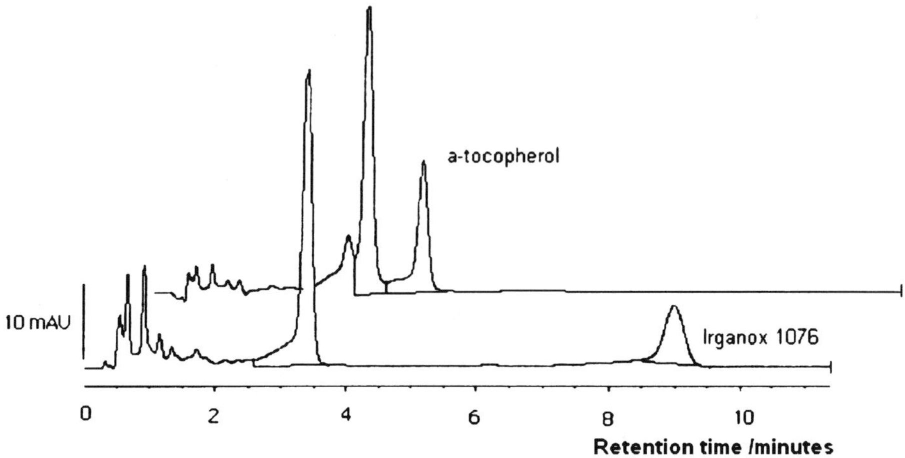

Fig. 4 HPLC chromatograms of residual antioxidant amount (%) of α-tocopherol (*top*) and Irganox 1076 (*below*), after MAE from un-aged film-blown PE. © owned by the authors

grinding). Albertsson et al. have in several studies developed methods for the quantitative determination of phenolic antioxidants in PE and PP by HPLC and MAE [17, 68, 69]. Figure 4 shows the HPLC chromatograms for two different antioxidants, α-tocopherol and Irganox 1076, extracted from film-blown PE by MAE. MAE was performed with ACN, and Irganox 1010 was used as an internal standard. In the HPLC analysis, the concentration of antioxidants was estimated from the absorption at 220 nm.

The residual antioxidant content (Table 3) was determined and compared to the stability of each material. In the material stabilized with α-tocopherol (PE-toc), 60% of the original amount of α-tocopherol was lost during the preparation of the films and during the film-blowing. In contrast, only 34% of the synthetic antioxidant Irganox 1076 (PE-Irg1076) was lost.

Table 3 Residual antioxidant content after processing in PE containing α-tocopherol (toc) or Irganox 1076 (Irg1076) and containing the additives oat starch (OS) or EAA [68]. The initial amount was 0.1 wt-%. © owned by the authors

Material	Residual antioxidant (wt-%)
PE-toc	0.040 ± 0.0015
PE-Irg1076	0.066 ± 0.0012
PE-EAA/toc	0.017 ± 0.0003
PE-EAA/Irg1076	0.066 ± 0.0012
PE-OS/toc	0.032 ± 0.0003
PE-OS/Irg1076	0.064 ± 0.0009

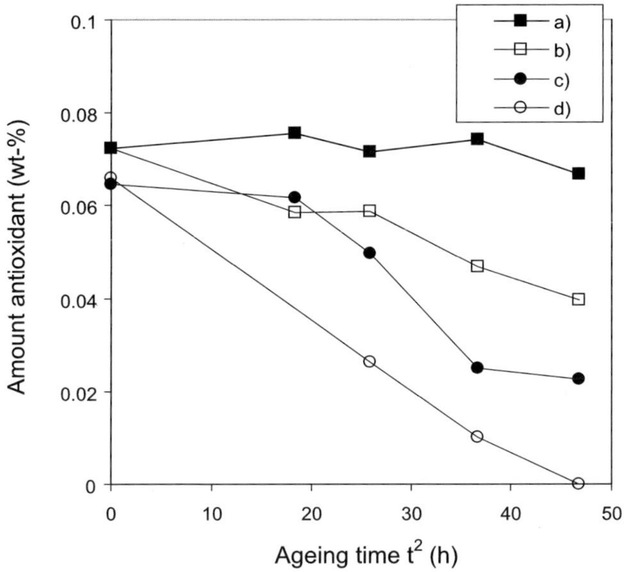

Fig. 5 Residual antioxidant amount (%) in PE (compression-molded) of Irganox 1076 aged in **a** O_2 (■), **b** MeOH/H_2O (□), of α-tocopherol aged in **c** O_2 (•), and in **d** MeOH/H_2O (○), determined by HPLC after MAE

Figure 5 shows the change in antioxidant content during ageing for up to 2200 h for compression-molded PE containing α-tocopherol (PE-toc_CM) and Irganox 1076 (PE-Irg1076_CM).

Lundbäck et al. [46] used HPLC and MAE to determine the amount of antioxidant Santonox R among others in PE samples aged in different environments. Figure 6 shows the loss of Santonox R, with a retention time of 1.2 min, during ageing in oxygen-free water at 90 °C. During the ageing an increase in peak height is visible for the peaks with a retention time between

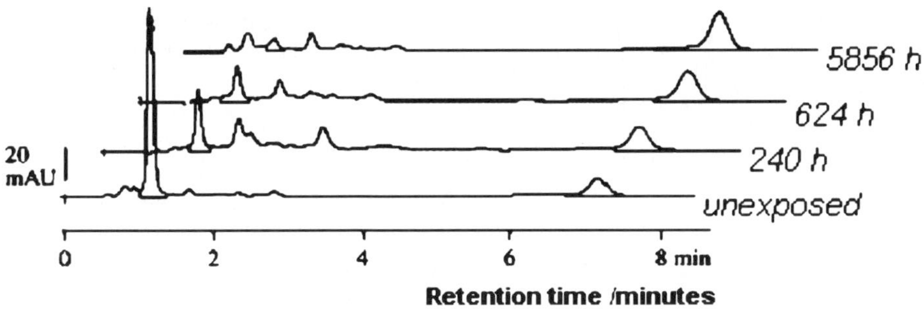

Fig. 6 HPLC chromatograms of Santonox R extracted by MAE from PE plates, un-aged and aged at different times in oxygen-free water at 90 °C. Santonox R peak at 1.6 min and internal standard peak at 7.3 min. © owned by the authors

1.6 and 2.8 min. These peaks are considered to originate from degradation products of the antioxidant.

Recently, Marcato and Vianello presented a large study on MAE of additives in polyalkenes [70]. Two MAE methods were reported, a one-step MAE (2.5 g polymer, 25 mL ethyl acetate : hexane 75 : 25, 125 °C, 15 min) useful for additives with low-medium polarity such as antioxidants, and

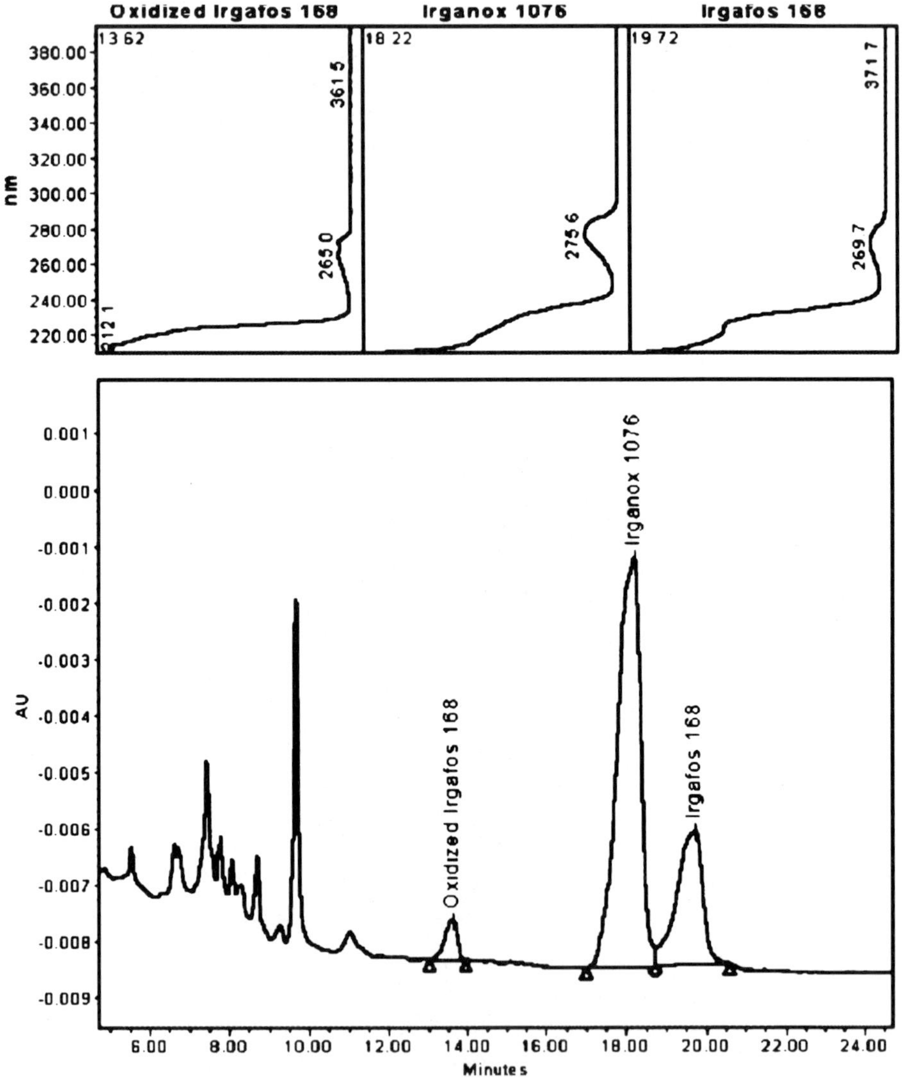

Fig. 7 LC chromatograms of Irgafos 168, Irganox 1076 and of the degradation product from Irgafos 168, extracted by MAE from PE. Permission for reproduction from Elsevier [71]

a MAE method including a manual shaking step. Both processes demonstrated excellent recovery. Dopico Garcia et al. [71] studied the quantification of Irganox 1076, Irgafos 168 and a degradation product from Irgafos 168, tri(2,4-di-*tert*-butylphenyl)phosphate in PE by MAE and reverse-phase LC. Extraction conditions were selected based on a low temperature and a short time to prevent degradation of Irgafos 168. DCM (30 mL) was shown to be the best solvent (the extraction was performed for 60 min at a temperature of 50 °C). Gradient LC with methanol : water was used. Figure 7 shows the chromatogram of the compounds studied.

Burman [72] has investigated the optimum MAE conditions of extraction of Irganox 1010 from PP films. Two different solvent combinations, isopropyl alcohol (IPA) : cyclohexane and acetone : cyclohexane were compared. Figures 8 and 9 show the extraction yields from these two combinations, re-

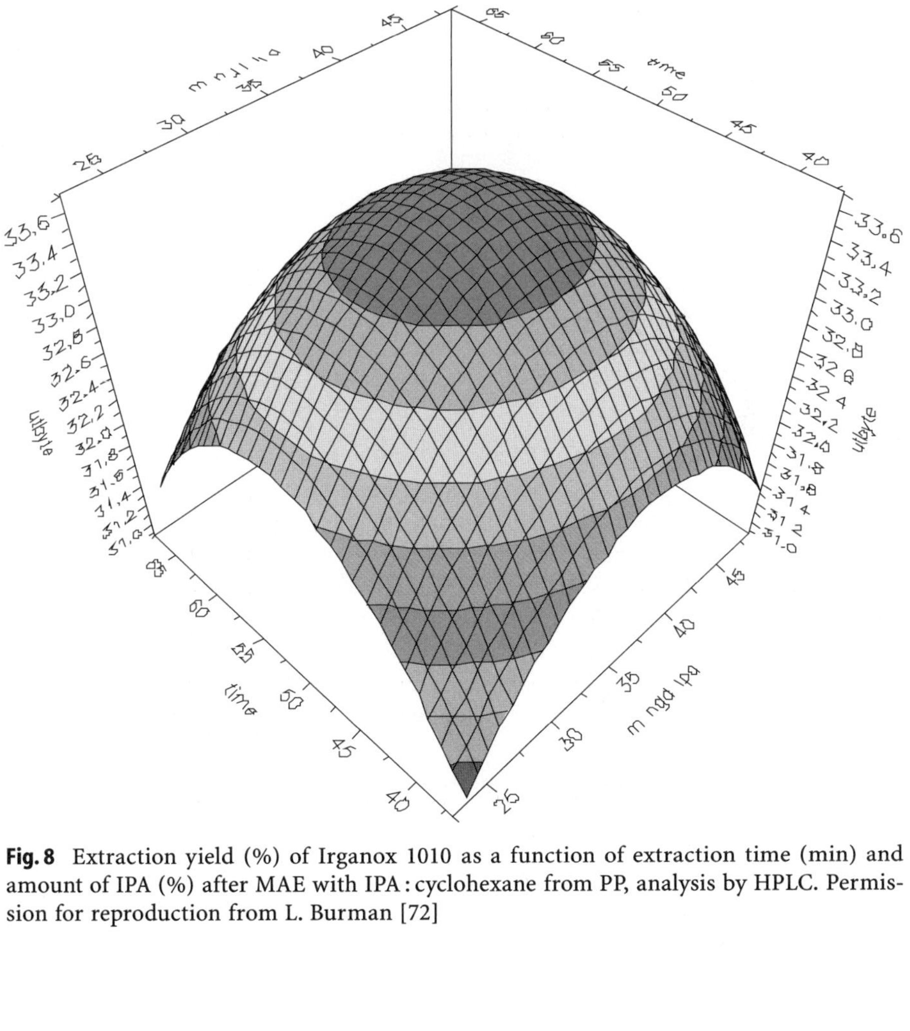

Fig. 8 Extraction yield (%) of Irganox 1010 as a function of extraction time (min) and amount of IPA (%) after MAE with IPA : cyclohexane from PP, analysis by HPLC. Permission for reproduction from L. Burman [72]

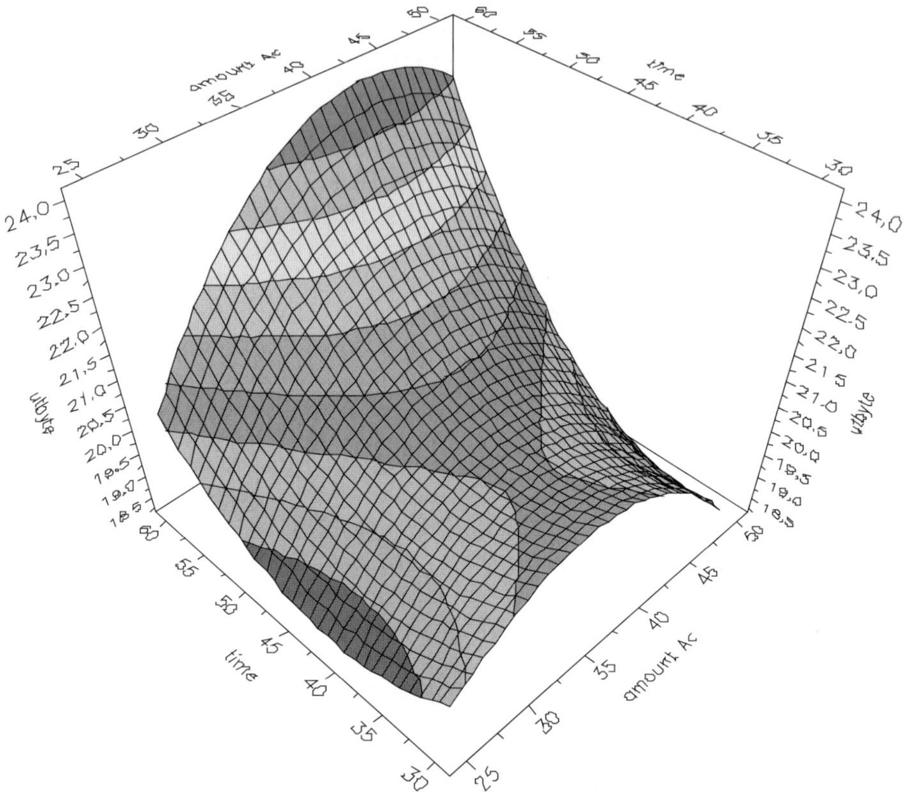

Fig. 9 Extraction yield (%) of Irganox 1010 as a function of extraction time (min) and amount of IPA (%) after MAE with IPA : acetone from PP, analysis by HPLC. Permission for reproduction from L. Burman [72]

spectively. The proportion of IPA or acetone was varied between 25 and 50 v/v-% (y-axis) and the time was varied between 30 and 60 min at a constant temperature of 60 °C. The yield is plotted on the z-axis. From the plots, the optimum yield was found. With IPA : cyclohexane as extracting solvent, the optimum yield of Irganox 1010 was 34% (40% IPA, 50 min extraction time at 60 °C), and with acetone : cyclohexane it was lower, only 24% (40% acetone, 60 min extraction time at 60 °C). The PP films were made by film-blowing, which is known to cause a loss of antioxidant.

It has been shown that reverse-phase HPLC is a more sensitive technique than normal-phase HPLC in detecting antioxidants [30], and that the use of IPA as extracting solvent gives high yields [73, 74]. In the study by Burman [72], not only non-polar solvents but also polar solvents, i.e. ACN and IPA, matched with the solvents used in reverse-phase HPLC, were investigated as extracting solvents (Tables 4 and 5). The choice of solvent had a greater impact on the yield than the temperature or the time, and the highest yield

Table 4 Extraction yields of Irganox 1010 from PP by MAE with different solvents, temperatures and times. Permission for reproduction from L. Burman [72]

Solvent Conditions	Extraction yields (%) 120 °C, 20 min	120 °C, 40 min	100 °C, 20 min	120 °C, 40 min	110 °C, 30 min
ACN	0.31 ± 0.01	0.32 ± 0.02	0.30 ± 0.01	0.33 ± 0.05	0.32 ± 0.03
IPA	0.36 ± 0.03	0.37 ± 0.03	0.39 ± 0.05	0.36 ± 0.02	0.40 ± 0.04
ACN : IPA (50 : 50)	0.30 ± 0.01	0.30 ± 0.02	–	–	–

Table 5 Extraction yields of Irgafos 168 from PP by MAE with different solvents, temperatures and times. Permission for reproduction from L. Burman [72]

Solvent conditions	Extraction yields (%) 120 °C, 20 min	120 °C, 40 min	100 °C, 20 min	120 °C, 40 min	110 °C, 30 min
ACN	1.11 ± 0.03	1.02 ± 0.05	1.00 ± 0.03	1.05 ± 0.12	1.00 ± 0.08
IPA	1.17 ± 0.08	0.86 ± 0.06	0.90 ± 0.11	0.86 ± 0.03	0.94 ± 0.08
ACN : IPA (50 : 50)	1.10 ± 0.02	0.94 ± 0.07	–	–	–

of Irganox 1010 was obtained with IPA as solvent. A higher temperature and a short time gave a high extraction yield of Irgafos 168, and the yield at the higher temperature was independent of the solvent used.

The RSD (Table 6) of MAE of Irganox 1010 and Irgafos 168 from PP was determined with IPA as solvent. Four extractions were carried out for each extraction. The RSD was found to be between 6 and 10%.

Table 6 RSD of MAE of Irganox 1010 and Irgafos 168 from PP, extracted at different temperatures and times with ACN. Permission for reproduction from L. Burman [72]

Antioxidant conditions	RSD (%) 120 °C, 20 min	110 °C, 30 min
Irganox 1010	6	10
Irgafos 168	10	6

3.3.3.2
Accelerated Solvent Extraction

An accelerated solvent extraction (ASE) is carried out under high pressure in a closed steel cell in an oven. The solvent is retained in the liquid phase at a temperature above its boiling point. This accelerates the extraction kinetics and forces the solvent into the polymer matrix [75, 76]. The pressure is usually around 10 MPa, the temperature between 100 and 200 °C, and the extraction time 15 min. A fully automatic system can be used for the extraction. Generally, a higher temperature and a solvent that acts as a good swelling agent for the polymers increase the extraction efficiency. The first selection of a suitable solvent can be difficult since literature data are available for most solvents only at normal pressure. The advantages compared to traditional techniques are several: shorter extraction time, low solvent consumption, good extraction yields and high reproducibility. Also, both aqueous and organic solvents can be used and this gives the method a wider possible choice of solvents than, e.g., Soxhlet extraction. The drawbacks are a high initial cost and less available information about this technique than about the other techniques.

ASE of Irganox 1076 from linear low-density PE (LLDPE) granules was performed with ethyl acetate : hexane (90 : 10) as extracting solution in 100 °C at 10.5 MPa, before and after radiation [77]. At first, THF was used as extracting solvent, but the polymer matrix melted at only 75 °C. The effect of altering the percentage of hexane was then studied. A higher percentage of hexane gave a higher yield of Irganox 1076, when the extraction was performed at a low temperature. Hexane may have acted as a swelling agent, especially at the lower temperatures. It was also observed that the temperature had a greater influence on the yield of Irganox 1076 than the proportion of hexane in the mixture with ethyl acetate. ASE was applied for the extraction of monomers and oligomers from nylon-6 and from poly(1,4-butylene terephthalate) (PBT) [78]. The effects of various experimental parameters, e.g. temperature, pressure, static time and flow rate, on the ASE extraction efficiency were investigated. It was found that the extraction efficiency of ASE depended primarily on the solvent but also on the extraction temperature. Vandenburg et al. [79] extracted Irganox 1010 from PP by ASE. The influences of temperature and of various solvents and solvent mixtures on swelling of the polymer were studied. With IPA as solvent, a 90% yield of Irganox 1010 was obtained from freeze-ground PP within 5 min at a temperature of 150 °C. Maximum extraction rates were found when a solvent or a solvent mixture was used at a temperature just below the melting point of the polymer. Suitable ASE conditions for extraction of BHA, Irganox MD 1024, BMP, Irgafos 126, HP 136, Irganox 3114, Tinuvin 328, Irganox 1010, Irganox 1330, Irganox 1076 and, Irgafos 168 from freeze-ground PE have been investigated [80]. IPA was used as solvent, at a pressure of 10.3 MPa. Different swelling agents, CH_x and THF (up to concentrations of 7.5%), extraction times, and temperatures (80–110 °C) were tested. The optimal conditions were obtained with 7.5% cyclohexane, at a temperature of 105 °C for a time of 15 min.

3.3.3.3 Supercritical Fluid Extraction

Supercritical fluid extraction (SFE) eliminates the use of large quantities of solvents. Carbon dioxide (CO_2) is usually used as solvent above its critical temperature and pressure [76]. It easily obtains a supercritical fluid state and it has low toxicity and low cost. The low viscosity and high diffusivity of the supercritical fluid with swelling of the polymer allows rapid extraction. SFE is suitable for thermally labile compounds (since relatively low temperatures can be used), it can be automated, is easily interfaced with SFC and it eliminates the use of organic solvents. However, the equipment is expensive and the optimization of the parameters is sometimes difficult. In the extraction of polar compounds a so-called polar chemical modifier, e.g. methanol, is added to increase the solubility of the antioxidants in the fluid. Thilen and Shishoo [34] optimized SFE for the extraction of Irganox 1010 and Irgafos 169 from PP followed by HPLC analysis. The temperature, the pressure, and modifiers were varied to find the best extraction conditions. Optimal extraction was achieved at a temperature of 120 °C and a pressure of 384 bar where the yield of Irganox 1010 was 62% and of Irgafos 168 74%. Garde et al. [33] extracted Irgafos 168, Irganox 1010 and Hostanox O3 by SFE from PP. The extraction efficiency for the smaller antioxidants was ca. 75% after a time of 90 min, but both Irganox 1010 and Hostanox O3 had a poorer recovery. SFE has been used for the extraction of Irganox 1010 and Irgafos 168 from PE [81]. It was seen that SFE was a selective extraction method, with efficient extraction of the antioxidants and that it was less sensitive to interference from low molecular weight polymer than Soxhlet extraction. By combining the FT-IR information for the extracts with SFC analysis, degradation products of Irgafos 168 could be detected. The temperature and the pressure, and also the thickness of the sample have been reported to have the greatest influence on the extraction efficiency in SFE of Irganox 1076, Irgafos 168 and Chimassorb 81 from PE [82].

3.3.3.4 Other Pressurized Techniques

Macko et al. [35] quantified thioether antioxidants (including Santonox R) in PE by HPLC, after extraction under pressure in an autoclave in 100 mL of *n*-heptane : IPA (97 : 3 v/v). The temperature was ramped up to 160 °C over 15 min and this temperature was then held for 75 min. Only about 80% of the original amount of Santonox R was recovered, presumably due to degradation of the antioxidant. In chromatograms of both the PE extract and a standard solution aged for 3 days, a small extra peak was seen. This indicated that some of the antioxidant was consumed during the extrusion of the samples, or that degradation of the substance occurred during the extraction, which would be a drawback of the method. As previously mentioned (Sect. 3.3.3.1), Lundbäck et al. [46] have seen an increasing abundance of a degradation product from Santonox R during thermal ageing.

3.3.4 Comparison Between Extraction Techniques

Table 7 shows a comparison between traditional and recent extraction techniques. In general, the MAE technique is easy to use and is cheaper than the other modern techniques, i.e. SFE and ASE. A drawback is that the samples have to be filtered after MAE. In SFE, clean-up is usually not needed because it is a relatively selective technique. However, method development is often more complex in SFE and another negative aspect is that the sample throughput is not as high as in MAE. In ASE, a filtration step is included in the technique but, as in MAE, a clean-up step is often needed. SFE and ASE do not give complete recovery of organic additives with high polarity [70], while MAE methods achieve good recoveries of these compounds in quite a short time.

Soxhlet extraction by methylene chloride (250 mL) of Irganox 1010 and of two phosphatic antioxidants (Irgafos 168 and Ultranox 626) from PP for 6 h at 50 °C has been shown to cause less degradation of the two phosphatic antioxidants than polymer dissolution precipitation at 100 °C after analysis by reverse-phase HPLC or MS [32]. The limit of detection was 0.002 wt-% for the three antioxidants and the RSD varied between 1.1 and 2%. Table 8 shows a comparison between Soxhlet extraction (1.5 g PE in 100 mL chloroform, for 5 h), MAE (0.3 g PE in 20 mL ACN, for 40 min), and UAE (0.3 g sample in 8 mL chloroform, for 1 h) of Santonox R, Irganox 1081 and Irganox 1010 from PE. The initial amount of antioxidant in the materials before processing was 0.1 wt %.

MAE of Santonox R from compression-molded PE plates (>100 μm thick) gave the same recovery as the Soxhlet extraction, and MAE of Irganox 1010 from thin films (ca. 30 μm thick) was more efficient than both Soxhlet and UAE. MAE of Irganox 1081 (also from PE plates) gave 17% less extracted antioxidant than the Soxhlet extraction. This shows that MAE is more dependent on sample thickness than Soxhlet extraction and that the choice of solvent had a smaller effect. Generally, the advantage of MAE in comparison to Soxhlet is that it is a faster method with less solvent usage. That MAE is as efficient as Soxhlet extraction has also been shown in another study [70], where Soxhlet and MAE were used for the determination of Irganox 1076

Table 7 Overview of traditional and more recent extraction methods

Parameters	Extraction method Soxhlet	Polymer dis.	UAE	MAE	SFE	ASE
Extraction time	2–48 h	60 min	40–60 min	3–30 min	20–120 min	5–30 min
Sample size (g)	1–5	>1	5–10	1–10	0.01–0.1	1–30
Solvent usage (mL)	50–100	>20	10–50	10–40	10	30–50
Filtration needed	–	–	–	x	–	x
Investment	Low	Low	Low	Moderate	High	High
Advantages	Inexpensive Widely accepted	Inexpensive	Inexpensive	Short ext. time	Low solv. use	Short ext. time
Drawbacks	Time-consuming High solvent use	Time-consuming	–	–	Expensive Difficult to optimize	Expensive

x Filtration needed

Table 8 MAE compared to Soxhlet and UAE of Santonox R and Irganox 1081 from compression-molded PE plates and of Irganox 1010 from PE films. Strandberg and Albertsson (unpublished data) and data by permission from L. Burman [72]

Extraction method	Residual content (%) Santonox R	Irganox 1081	Irganox 1010 [a]
Soxhlet	92±0.5	97.1±1.1	28±2
MAE	94±2	80.2±2.3	36±4
UAE	–	–	29±3

[a] Extruded PE film

in PE. MAE with ACN was found to give approximately the same yield as a Soxhlet extraction with chloroform, and ACN could thus be used as extracting solvent, sample solvent and mobile phase in the HPLC analysis. This simplified the analysis process. The mass limit of detection of the method was found to be 3.3 ng (a concentration limit of detection of 33 ng/mL). The original amount of antioxidant was 0.6 wt-%, and the RSD was less than 1.2% [73]. The extraction of low molecular mass oligomers and cyclic trimers from PET film has been investigated by Costley et al. [83]. At optimized MAE conditions (8 g PET in 40 mL DCM, at 120 °C) good recovery was obtained after only 120 min. At temperatures over 120 °C, the polymer fused. Conventionally, antioxidants in PET have been Soxhlet extracted with xylene for 24 h. Smith and Taylor [84] compared SFE to the traditional polymer dissolution process for the extraction of Irganox 1076, Tinuvin P, Tinuvin 770 and Wytox from PS. Only a small difference in extraction yield could be seen between the two methods, and the SFE had a RSD ranging from 2 to 13%.

The extraction efficiency of MAE and ASE has been compared to that of UAE and to traditional techniques (i.e. Soxhlet extraction, polymer dissolution and shake-flask extraction) in extracting Irganox 1010 from 0.2 to 0.3 g freeze-ground PP [74]. Among the UAE and the traditional techniques, polymer dissolution gave the highest extraction efficiency (99%) with chloroform (30 mL) as extracting solvent after a time of 60 min. ASE and MAE gave significantly faster extractions with the same recoveries as the polymer dissolution. The ASE needed a time of only 10 min when IPA was used as extracting solvent at 150 °C. MAE gave a high recovery (96%) after only 5 min of extraction in IPA (30 mL) at 150 °C. SFE and enhanced fluid extraction (ESE) have been compared to a polymer dissolution method in extracting Irganox 1076, Irgafos 168 and Irganox 1010 from PP [85]. The extraction yields of Irganox 1076 and Irgafos 168, after the SFE or ESE were comparable to those obtained after polymer dissolution, whereas for Irganox 1010 the polymer dissolution method gave a higher yield.

3.4 Relation Between Residual Antioxidant Content, Build-up of Degradation Products and Stability in Polyolefins

3.4.1 General Aspects

Determination of the residual antioxidant content in polymers by HPLC and MAE is one way to determine the amount needed for reasonable stabilization of a material, and also to compare different antioxidants and their individual efficiencies. During ageing and oxidation of PE, carboxylic acids, dicarboxylic acids, alcohols, ketones, aldehydes, *n*-alkanes and 1-alkenes are formed [86–89]. The carboxylic acids are formed as a result of various reactions of alkoxy or peroxy radicals [90]. The oxidation of polyolefins is generally monitored by various analytical techniques. GC-MS analysis in combination with a selective extraction method is used to determine degradation products in plastics. FTIR enables the increase in carbonyls on a polymer chain, from carboxylic acids, dicarboxylic acids, aldehydes, and ketones, to be monitored. It is regarded as one of the most definite spectroscopic methods for the quantification and identification of oxidation in materials, and it is used to quantify the oxidation of polymers [91–95]. Mechanical testing is a way to determine properties such as strength, stiffness and strain at break of polymeric materials.

3.4.2 Antioxidant Content, Carbonyl and Carboxylic Acid Build-up in Stabilized Polyethylene

Quantification by FTIR of the whole compound classes of carbonyl-containing degradation products formed in stabilized PE during aging has been compared to the remaining antioxidant content in the materials determined by MAE and HPLC analysis (Table 3) [17, 68]. Figure 10 shows the carbonyl index for PE stabilized with α-tocopherol (PE-toc) and for PE stabilized with Irganox 1076 (PE-Irg1076), as a function of the ageing time. PE-toc showed a faster increase in the carbonyl index than PE-Irg1076 during the ageing, and the PE-Irg1076 material had a longer induction time. This agrees with the observation that less antioxidant was left in PE-toc than in PE-Irg1076 after the film-blowing (Table 3), leading to a decreased prevention of oxidation in this material.

For the quantification of only one compound (e.g. propanoic acid) rather than of the whole class of compound (e.g. carboxylic acids), more precise techniques such as mass spectrometry (MS) are demanded, together with a calibration curve based on the peak areas of suitable standard compounds as a function of their concentration. The major challenge is then to selec-

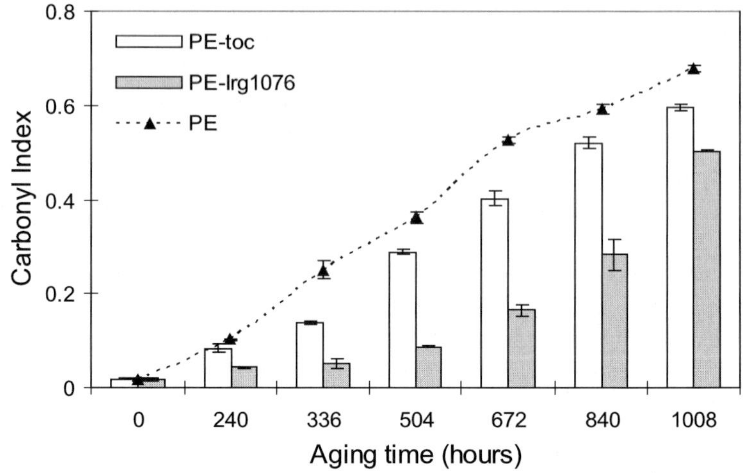

Fig. 10 Increase in carbonyl index during aging in air at 80 °C for film-blown PE: un-stabilized (▲) (PE), containing α-tocopherol (PE-toc), and containing Irganox 1076 (PE-Irg1076), measured by FTIR in ATR mode. © owned by the authors

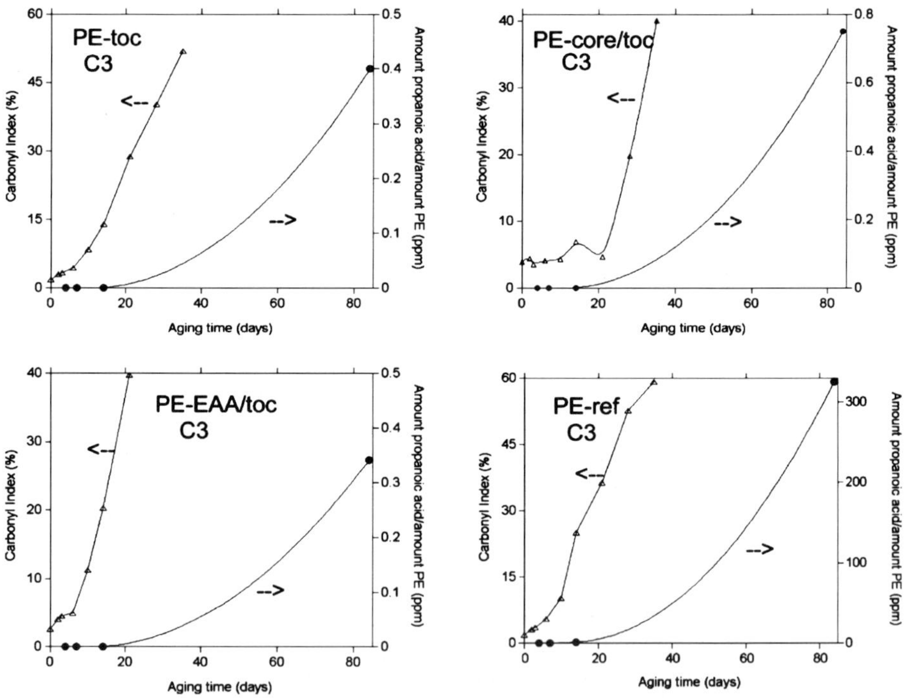

Fig. 11 Carbonyl index (△) (determined by ATR FTIR) and emitted amount of propanoic acid (•) (extracted by HS-SPME and identified by GC-MS) in un-stabilized PE (PE-ref) and in PE materials containing α-tocopherol: PE-toc, PE-core/toc and PE-EAA/toc, during aging in air at 80 °C. The authors have the right to republish [45]

tively extract degradation products (present only in low concentrations) from the solid polymer matrix or from, e.g., an aqueous extract. The extraction can be performed by, e.g., solid-phase extraction (SPE) or by solid-phase microextraction (SPME) [96]. In a recent study, the carbonyl index in solid PE was compared with each of the carboxylic acids in the homologous series from propanoic to decanoic acid during ageing [97]. The amount of the carboxylic acids was assessed by headspace-SPME (HS-SPME) and GC-MS. The analytes were extracted from closed vials containing PE samples aged up to 84 days. The relations between the build-up of propanoic acid and the carbonyl index in un-stabilized PE (PE-ref), PE-toc, PE containing α-tocopherol and poly(ethylene-*co*-acrylic acid) (PE-EAA/toc), and PE containing α-tocopherol and core-shell polymers (PE-core/toc) are shown in Fig. 11.

3.4.3
Antioxidant Content and Mechanical Properties in Polybutadiene

Celina et al. [98] have recently investigated the correlation between the content of the antioxidant 2,2′-methylene-bis-(4-methyl-6-*tert*-butylphenol) in polybutadiene and the mechanical performance during aging at tempera-

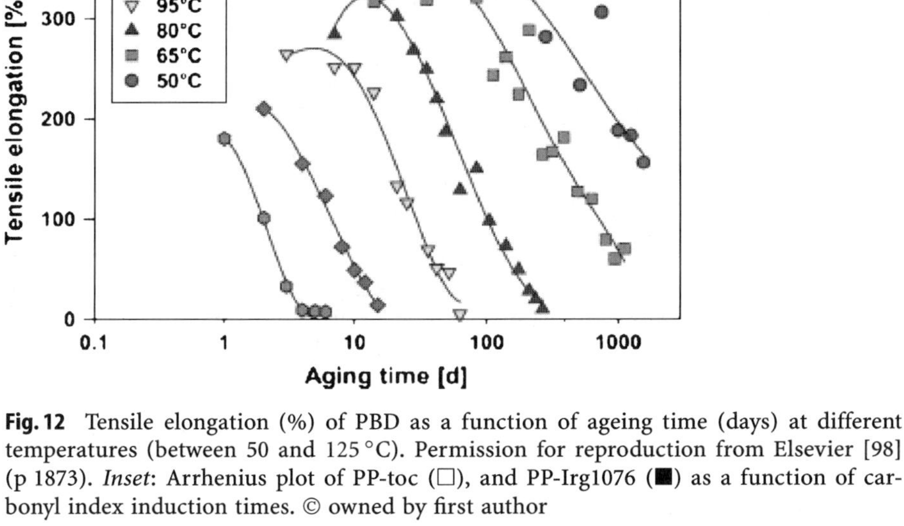

Fig. 12 Tensile elongation (%) of PBD as a function of ageing time (days) at different temperatures (between 50 and 125 °C). Permission for reproduction from Elsevier [98] (p 1873). *Inset*: Arrhenius plot of PP-toc (□), and PP-Irg1076 (■) as a function of carbonyl index induction times. © owned by first author

tures ranging from 50 to 125 °C. The antioxidant content was determined by GC after extraction in xylene for 16 h at room temperature. Degradation products from the antioxidants (e.g. di- or trimer degradation products of the antioxidant) were not found, presumably because of too-low temperatures in the GC analysis. It was seen that at higher aging temperatures, e.g. 80–110 °C, the content of antioxidant decreased more rapidly than the mechanical properties, i.e. tensile elongation. At the lower aging temperatures, 50 °C or 65 °C, the antioxidant content decreased more slowly with aging time than the change in tensile elongation. Even so, the mechanical degradation was not prevented (Fig. 12).

The antioxidant depletion had an activation energy (E_a) of about 135 kJ/mol at the higher and of about 65 kJ/mol at the lower temperatures. It is well known that a non-Arrhenius behavior is observed for a material aged at different temperatures, so that a linear extrapolation of the induction time from a higher to a lower temperature is misleading, and that an extrapolated induction time will be longer than the true induction time [100, 101]. Strandberg and Albertsson [99] have earlier shown that stabilized PP subjected to accelerated aging at temperatures of 60–120 °C shows non-Arrhenius behavior (Fig. 12, inset graph) since the temperature affects the polymer morphology, the solubility and migration of additives, and the mechanism of antioxidant loss [102].

4 Leaching of Antioxidants and Their Degradation Products Into Solution

4.1 General Concerns

Plastic materials are widely used in medical and food applications, in containers, tubing and special devices. Generally, plastics are considered to be chemically inert but the migration of substances from plastics into products is well known. Hence, the use of plastics in contact with products must be controlled by the identification of migrating compounds and the quantification of their accumulation levels. Migration is mostly due to the low solubility of the compounds in the polymer [9], and is also known as leaching. According to the European Union legislation, components in plastics in contact with food should have a migration level under certain defined threshold values [103]. Low molecular mass compounds, such as additives and their degradation products, have been seen to migrate from polymers into the surrounding environment [104–107] and accumulate there. This may affect the viability of the product and cause off-flavors [108] and toxicity [10].

4.2 Migration of Antioxidants and Their Degradation Products From Polymers Into Solution

4.2.1 Migration Into Food or Aqueous Simulants

Various procedures have been used to study the migration levels of antioxidants by chromatography. Till et al. [109] used HPLC to measure the migration of BHT from HDPE into food and food simulants. As expected, the migration of the antioxidant into oil and fatty foods was more rapid than the migration into aqueous solutions. Anyhow, the diffusion of the antioxidant within the polymer was seen to be rate-limiting for the migration. Lawson et al. [106] have used GC-MS to study the migration of the Irganox and Irgafos antioxidants from food packaging at an elevated temperature. The levels of migration of the antioxidants were lower than the officially permitted limits. Degradation products from the antioxidants were also detected. Garde et al. [110, 111] characterized the migration of Irgafos 168 and Irganox 1076 from PP into both fatty and aqueous food simulants. The aqueous simulants were fully evaporated with nitrogen at room temperature and the residues were dissolved in chloroform and analyzed by GC. Marque et al. [49] studied the migration of BHT, Irgafos 168, Irganox 1010, Irganox 1076 and their degradation products into a fatty food stimulant (isooctane) from a five-layered material, with PP in the contact layer, after irradiation of the plastic. Only BHT and the degradation products DBP (from Irgfos 168) and DBB were found to migrate into the isooctane. Berg et al. [112] studied migration of Irganox 1010 and Irgafos 168 from PP into an acid-based food stimulant (water containing acetic acid) by LLE with chloroform, followed by analysis by SFC-FID. In addition, Irganox 1010 was derivatized with acetic anhydride and analyzed by GC-MS. Dopico-Garcia et al. [113] used LLE to extract several antioxidants, including Irganox 1010 and Irganox 1076, from water. The yields obtained after a re-dissolution of the antioxidants are shown in Table 9. The optimum extraction conditions of the LLE were 20 mL *n*-hexane for 6 min at a pH of 0.5. Table 10 shows the yield of the antioxidants after the LLE extractions. The yields were quite low, especially for Irganox 1010. The method had a limit of detection between 16 and 30 $\mu g\,L^{-1}$. The same authors [113] reported a migration test [103] of the antioxidants Ethanox 330, Irganox 1010, Irganox 1076, BHT, BHA, and of Irgafos 168 and of one of its degradation products (DBP). Only Ethanox 330 and Irganox 168 could be detected. Higher recoveries and a lower RSD (Table 11) have been reported when using the solid-phase extraction of antioxidants from aqueous food simulants [114].

The hydrolytic stability of two antioxidants, Hostanox O3 and Hostanox O10, has been investigated by HPLC-ESI-MS [107]. The antioxidants were

Table 9 Extraction yields and RSD of different phenolic antioxidants after LLE, with different solvents, from water. Permission for reproduction from Elsevier [113]

Compound	Extracting solvent THF Yield (%)	RSD (%)	ACN Yield (%)	RSD (%)
BHT	76	13	82	9
DBP	85	7	86	9
BHA	85	5	86	7.5
Irganox 1010	83	6	88	8
Ethanox 330	83	5	86	7
Irgafos 168	88	10	86	10
Irganox 1076	91	12	86	7

Table 10 Extraction yield, RSD, limits of detection (LOD) and limits of quantification (LOQ) after LLE of aqueous samples analyzed by HPLC and UV ($n = 6$). Permission for reproduction from Elsevier [113]

Compound	Yield (%)	RSD (%)	LOD ($\mu g\ L^{-1}$)	LOQ ($\mu g\ L^{-1}$)
BHT	70	4	7	22
DBP	88	3	5	16
BHA	88	9	6	19
Irganox 1010	66	15	6	20
Ethanox 330	71	7	4	19
Irgafos 168	76	9	9	30
Irganox 1076	71	13	8	25

Table 11 Extraction yield, RSD, limits of detection (LOD) and limits of quantification (LOQ) of different phenolic antioxidants after SPE of aqueous samples (3% acetic acid and 10% ethanol) ($n = 7$) analyzed by LC. Permission for reproduction from Elsevier [114]

Compound	Yield (%)	RSD (%)	LOD ($\mu g\ L^{-1}$)	LOQ ($\mu g\ L^{-1}$)
BHA	104	2.0	0.052	0.17
DBP	97	3.0	0.038	0.13
BHT	82	6.2	0.045	0.15
Irganox 1010	83	6.6	0.048	0.16
Ethanox 330	80	6.6	0.034	0.15
Irgafos 168	86	7.7	0.071	0.24
Irganox 1076	78	6.5	0.060	0.20

Table 12 Amount of Irganox 1010 after 10 days, at 40 °C in different solutions containing HDPE; extraction by LLE and identification by TLC and LSC. Permission for reproduction from Springer [115]

Solvent	Amount of migrated antioxidant (%)	
	Thin-layer chrom	Liquid-scintillation count
Milk	0.6 ± 0.20	0.6 ± 0.07
Water–ethanol	–	0.2 ± 0.02
Olive oil	0.9 ± 0.10	1.1 ± 0.03

aged in water for one year, and it was found that the Hostanox O3 had a higher hydrolytic stability than the Hostanox O10. The leaching of antioxidants from plastic packaging into dairy products has been investigated by Haesen et al. [115] using LLE followed by thin-layer chromatography (TLC) or by liquid-scintillation counting. Carbon-14 labeling of one of the antioxidants showed that the migration of the antioxidant depended on the type of extracting solvent. As expected, a greater migration was seen when the plastic was in contact with olive oil than with milk (Table 12).

4.2.2 Migration Into Pharmaceutical Solution

Burman and Albertsson [116] have developed methods for quantifying certain theoretical degradation products of Irganox 1010 and Irgafos 168 from pharmaceutical solutions consisting of 10% ethanol in water. The degradation products of Irganox 1010 were 2,4-di-*tert*-butylphenol, 2,6-di-*tert*-butyl-*p*-benzoquinone, 3,5-di-*tert*-butyl-4-hydroxyphenylpropionic acid, 2,6-

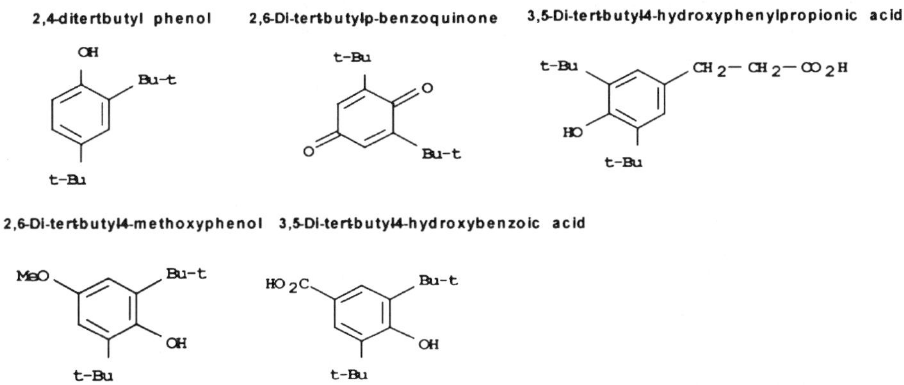

Scheme 7 Studied degradation products from Irganox 1010

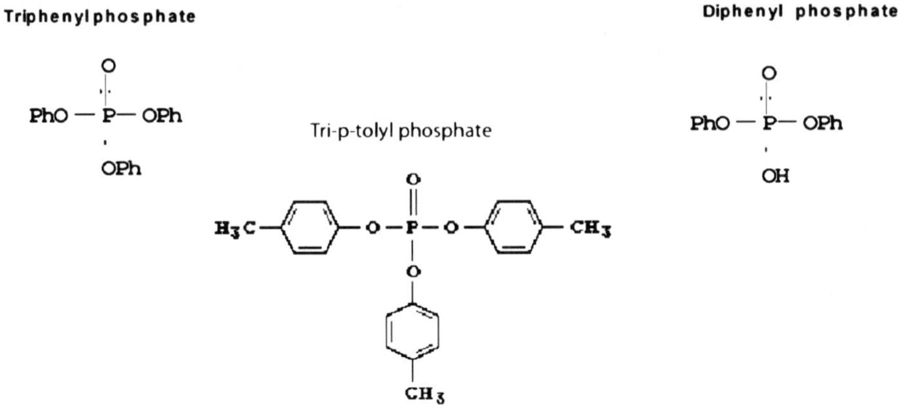

Scheme 8 Studied degradation products from Irgafos 168

di-*tert*-butyl-4-methoxyphenol and 3,5-di-*tert*-butyl-4-hydroxybenzoic acid (Scheme 7); and of Irgafos 168 were diphenyl phosphate, triphenyl phosphate and tri-*p*-tolyl phosphate (Scheme 8).

The PP samples were aged in glass vials at 80 °C for 4 months, and the degradation products were extracted by HS-SPME or by immersion SPME. The compounds extracted by HS-SPME were identified by GC-MS,

Table 13 Relative response, RSD and determined concentrations of degradation products of Irganox 1010 and Irgafos 168 in solutions containing PP samples, after HS-SPME. Permission for reproduction from Elsevier [116]

Compound	Response AU	RSD (%)	Response IS	RSD (%)	Concentration (μg L^{-1})
60 *min* at 55 °*C*					
2,6-Di-*tert*-butyl-*p*-benzoquinone	80 335 061	3	97.4	1	–
2,4-Di-*tert*-butylphenol	34 793 025	4	42.2	8	6
3,5-Di-*tert*-butyl-4-hydroxy phenylpropionic acid	665 463	4	0.8	8	12
3,5-Di-*tert*-butyl-4-hydroxybenzoic acid	824 830	4	–	–	–
60 *min* at 70 °*C*					
2,6-Di-*tert*-butyl-*p*-benzoquinone	32 909 432	4	3.4	7	–
2,4-Di-*tert*-butylphenol	24 951 350	1	2.6	4	4
3,5-Di-*tert*-butyl-4-hydroxy phenylpropionic acid	3 531 074	4	0.4	1	12
3,5-Di-*tert*-butyl-4-hydroxybenzoic acid	9 564 468	3	–	–	–

Table 14 Relative response, RSD and determined concentrations of the degradation products of Irganox 1010 and Irgafos 168 in solutions containing PP samples, after immersed SPME. Permission for reproduction from Elsevier [116]

Compound	Response AU	RSD (%)	Response IS	RSD (%)	Concentration ($\mu g\,L^{-1}$)
2,6-Di-*tert*-butyl-*p*-benzoquinone	22 229	8	0.17	9	–
2,4-Di-*tert*-butylphenol	6687	20	0.05	10	5
3,5-Di-*tert*-butyl-4-hydroxy phenylpropionic acid	60 033	11	0.46	6	53

whereas the compounds extracted by immersion SPME were identified by GC. A fiber with a polydimethylsiloxane–divinylbenzene (PDMS-DVB) coating was found to be the most suitable fiber for both the HS-SPME and the immersion SPME. A larger number of degradation products could be quantified simultaneously by the use of HS-SPME than by immersion SPME (Tables 13 and 14). Both ethylated and un-ethylated phenolic acids were detected by immersion SPME, and this technique had a higher sensitivity in the detection of quinone and 2,4-di-*tert*-butylphenol than HS-SPME. Nevertheless, at a low extraction temperature (55 °C) the HS-SPME could also be used for quantification of 2,4-di-*tert*-butylphenol.

5 Concluding Remarks

In this review we have summarized the results obtained by different chromatographic techniques and a variety of sample preparation methods for the analysis of antioxidants in polymers and in solutions. Efficient techniques including liquid and gas chromatography, mass spectrometry, traditional low pressure extraction techniques and newer high pressure techniques have been developed. These have made possible the accurate quantification and identification of antioxidants. The newer techniques offer versatile tools for further developments in this area of polymer analysis.

Microwave assisted extraction (MAE), supercritical fluid extraction (SFE) and accelerated solvent extraction (ASE) have made the extraction of thermally labile antioxidants possible. With ground films or thin films, MAE has been shown to have a high extraction efficiency with short extraction times and low solvent consumption, giving cost-effective methods. SFE has also been shown to give high extraction yields. There are drawbacks with the high pressure techniques; the optimization of the extraction parameters can be difficult and a greater investment in laboratory equipment is needed.

The development of analytical techniques for the identification and quantification measurement of antioxidants has accelerated, mainly because of their important stabilizing function in polymers and the related demand for production control and an increased knowledge of their migration out of plastics into food etc. The determination of residual antioxidant content in plastic materials, e.g. by high-performance liquid chromatography (HPLC) and extraction by MAE, has been shown to be a sophisticated technique for predicting long-term stability of polyethylene (PE). The relation between the antioxidant content in polybutadiene and the mechanical properties has been seen to have a dependence on the ageing temperature to which the material has been subjected. Using gas chromatography–mass spectrometry (GC-MS) and headspace–solid phase microextraction (HS-SPME) it has been shown that the antioxidant efficiency and build-up of degradation products in linear low density PE are related, especially for propanoic acid.

With low detection limit techniques, the identity, migration levels and accumulation of antioxidants and their degradation products in food, in liquids and in pharmaceutical solutions can be controlled. HPLC or GC-MS offers functional methods together with, e.g., liquid–liquid extraction (LLE). Recently, solid phase extraction (SPE) has been used successfully with lower detection limits and smaller errors than LLE. SPE is a technique that uses a small amount of solvent and that efficiently concentrates solutions of analytes depending on, e.g., polarity.

References

1. Taylor JS (1986) In: Fisher A (ed) Contact dermatitis. Lea & Febiger, Philadelphia, PA, p 603
2. Egsgaard H, Larsen E, Pedersen B, Carlsen L (1992) Trends Anal Chem 11:164
3. Albertsson A-C, Groening M, Hakkarainen M (2005) Polym Prepr 46:247
4. Albertsson A-C, Karlsson S (1990) Polym Mater Sci Eng 62:976
5. Karlsson S, Banhidi ZG, Albertsson A-C (1989) Mater Struct 22:163
6. Ezquerro O, Pons B, Tena MT (2003) J Chromatogr A 1008:123
7. Albertsson A-C, Groening M, Hakkarainen M (2003) J Appl Polym Sci 89:867
8. Stivala SS, Kimura J, Gabbay SM (1983) In: Allen NS (ed) Degradation and stabilisation of polyolefins. Elsevier Applied Science, London, p 111
9. Foldes E, Lohmeijer J (1997) J Appl Polym Sci 65:761
10. Foldes E (1998) Angew Makromol Chem 262:65
11. Foldes E (1995) Polym Degrad Stab 49:57
12. Kulich DM, Wolkowicz MD (1989) Adv Chem Ser 222:329
13. Wheeler DA (1968) Talanta 15:1315
14. Coe GR (1977) Soc Plast Eng Tech Pap 23:496
15. Spell HL, Eddy RD (1960) Anal Chem 32:1811
16. Drushel HV, Sommers AL (1964) Anal Chem 36:836
17. Strandberg C, Albertsson A-C (2005) J Appl Polym Sci 98:2427
18. Freitag JF, Fresenius Z (1983) Anal Chem 316:495

19. Schroeder HF, Zeynalov EB, Bahr H, Rybak T (2002) Polym Polym Compos 10:73
20. Brandolini AJ, Garcia JM, Truitt RE (1992) Spectroscopy 7:34
21. Bellamy LJ, Lawrie JH, Press EWS (1947) Trans Inst Rubber Ind 23:15
22. Heide RF, Wouters O (1962) Z Lebensm Untersuch Forsch 117:129
23. Thomas MP (1996) J Vinyl Add Tech 2:330
24. Larsen FN (1969) Appl Polym Sup 8:111
25. Howard IM (1971) J Chromatogr 55:15
26. Coupek J, Pokorny S, Protivova J, Holcik J, Karvas M, Pospisil J (1972) J Chromatogr 65:279
27. Wims AM, Swarin S (1975) J Appl Polym Sci 19:1243
28. Majors RE (1970) J Chromatogr Sci 8:339
29. Munteanu D, Isfan A, Isfan C, Tincul I (1987) Chromatographia 23:7
30. Nielson RC (1993) J Liq Chromatogr 16:1625
31. Nielson RC (1991) J Liq Chromatogr 14:503
32. Mansouri HE, Yagoubi N, Ferrier D (1998) Chromatographia 48:491
33. Garde JA, Catalá PR, Vergara R (1998) Food Addit Contam 15:701
34. Thilen M, Shishoo R (2000) J Appl Polym Sci 76:938
35. Macko T, Furtner B, Lederer K (1997) Int J Polym Anal Char 3:369
36. Schabron JF, Fenska LE (1980) Anal Chem 52:1411
37. Bruheim I, Molander P, Lundanes E, Greibrokk T, Ommundsen E (2000) J High Resolut Chromatogr 23(9):525–530
38. Somsen GW, Rozendom EJE, Gooijer C, Velthorst NH, Brinkman UAT (1996) Analyst 121:1069
39. Nerin C, Salafranca J, Cacho J, Rubio C (1995) J Chromatogr A 690:230–236
40. Crozier A, Jensen E, Lean MEJ, McDonald MS (1997) J Chromatogr A 761:315
41. Priha E, Ahonen L (1998) J Occupational Health 40:232
42. Hill SS, Shaw BR, Wu AHB (2003) Biomed Chromatogr 17:250
43. Raynor MW, Bartle KD, Davies IL, Williams A, Clifford AA, Chalmers JM, Cook BW (1988) Anal Chem 60:427
44. Gedde WU (1995) Polymer Physics. Chapman & Hall, London, p 232
45. Strandberg C (2006) Long-Term properties of polyethylene films – efficiency of a natural antioxidant. Thesis, Royal Institute of Technology (KTH), Stockholm, Sweden
46. Lundbäck M, Strandberg C, Albertsson AC, Hedenqvist MS, Gedde UW (2005) Polym Degrad Stab 91:1071
47. Vandenburg HJ, Clifford AA, Bartle KD, Carroll J, Newton I, Garden LM, Dean JR, Costley CT (1997) Analyst 122:101R
48. Luque de Castro MD, Garcia-Ayuso LE (1998) Anal Chim Acta 369:1
49. Marque D, Feigembaun A, Dainelli D, Riquett A-M (1998) Food Addit Contam 15:831
50. Haney MA, Dark WA (1980) 18:655
51. Di Pasquale M, Galli M (1984) High Resolut Chromatogr 7:484
52. Crompton TR (1968) Eur Polym J 4:473
53. Lichtenthaler RG, Ranfelt F (1978) J Chromatogr 149:553
54. Garde JA, Galotto J, Catala R, Gavara R (1998) ISLI Symposium on food packaging: ensuring the quality and safety of foods, Budapest, Hungary, 11–13 Sept 1996. Food Addit Contam 15:701
55. Perlstein P (1983) Anal Chim Acta 149:21
56. Al-Malaika S, Ashley H, Issenhuth S (1994) J Polym Sci Part A: Polym Chem 32:3099
57. Al-Malaika S, Goodwin C, Issenhuth S, Burdick D (1999) Polym Degrad Stab 64:145
58. Al-Malaika S, Issenhuth S (1999) Polym Degrad Stab 65:143
59. Al-Malaika S, Issenhuth S (2001) Polymer 42:2915

60. Al-Malaika S, Goodwin C, Issenhuth S, Burdick D (2001) Polym Degrad Stab 73:491
61. British Standard (1965) Code 2782
62. Schabron JF, Smith VJ, Ware JL (1982) J Liq Chromatogr 5:613
63. Francis VC, Sharma YN, Bhardwaj IS (1983) Angew Makromol Chem 113:329
64. Allen DW, Clench MR, Crowson A, Leathard DA (1993) J Chromatogr 629:283
65. Luque-Garcia JL, Luque de Castro MD (2003) Trends Anal Chem 22:4
66. Haider N, Karlsson S (1999) Analyst 124:797
67. Freitag W, John O (1990) Angew Makromol Chem 175:181
68. Strandberg C, Albertsson A-C (2006) J Polym Sci Part A: Polym Chem 44:1660
69. Burman L, Albertsson A-C (2005) J Polym Sci Part A: Polym Chem 43:4537
70. Marcato B, Vianello M (2000) J Chromatogr A 869:285
71. Dopico Garcia MS, Lopez VJM, Bouza R, Abad MJ, Gonzalez Soto E, Gonzalez Rodriguez MV (2004) Anal Chim Acta 521:179
72. Burman L (2005) Chromatography and extraction techniques for new evaluation methods of polyolefins long-term performance. Thesis, Polymer Technology, Royal Institute of Technology (KTH), Stockholm, Sweden, p 86
73. Molander P, Haugland K, Hegna DR, Ommundsen E, Lundanes E, Greibrokk T (1999) J Chromatogr A 864:103
74. Vandenburg HJ, Clifford AA, Bartle KD, Carroll J, Newton ID (1999) Analyst 124:397
75. Kaufmann B, Christen P (2002) Phytochem Anal 13:105
76. Raynie ER (2004) Anal Chem 76:4659
77. Waldeback M, Jansson C, Senorans FJ, Markides KE (1998) Analyst 123:1205
78. Lou XW, Janssen HG, Cramers CA (1997) Anal Chem 69:1598
79. Vandenburg HJ, Clifford AA, Bartle KD, Zhu SA, Carroll JI, Newton D, Garden LM (1998) Anal Chem 70:1943
80. Garrido-Lopez A, Tena MT (2005) J Chromatogr A 1099:75
81. Wieboldt RC, Kempfert KD, Dalrymple DL (1990) Appl Spectroscopy 44:1028
82. Salafranca J, Cacho J, Nerin C (1999) High Resolut Chromatogr 22:553
83. Costley CT, Dean JR, Newton I, Carroll J (1997) Anal Commun 34:89
84. Smith SH, Taylor LT (2002) Chromatographia 56:165
85. De Paepe A, Erlandsson B, Ostelius J, Gasslander U, Arbin A (2006) J Liq Chromatogr Rel Tech 29:1541
86. Albertsson AC, Barenstedt C, Karlsson S (1994) Acta Polym 45:97
87. Albertsson AC, Barenstedt C, Karlsson S (1995) J Chromatogr A 690:207
88. Karlsson S, Hakkarainen M, Albertsson A-C (1997) Macromolecules 30:7721
89. Hakkarainen M, Albertsson A-C, Karlsson S (1997) J Appl Polym Sci 66:959
90. Lacoste J, Carlsson DJ, Falicki S, Wiles DM (1991) Polym Degrad Stab 34:309
91. Matveeva EN, Khin'kis SS, Tsvetkova AI, Balandina VA (1963) Plast Massy 1:2
92. Adams JH (1970) J Polym Sci Polym Chem Ed 8:1077
93. Jadrnicek B, Stivala SS, Reich L (1970) J Appl Polym Sci 14:2537
94. Fodor Z, Iring M, Tudos F, Kelen T (1984) J Polym Sci Polym Chem Ed 22:2539
95. Cheng HN, Schilling FC, Bovey FA (1976) Macromolecules 9:363
96. Abalos M, Bayona JM, Pawliszyn J (2000) J Chromatogr A 873:107
97. Strandberg C, Albertsson A-C (2006) J Polym Sci: Part A Polym Chem 45:1848
98. Celina M, Skutnik Elliot JM, Winters ST, Assink RA, Minier LM (2006) Polym Degrad Stab 91:1870
99. Strandberg C, Albertsson A-C (2006) Eur Polym J 42:1855
100. Tamblyn JW, Newland GC (1965) J Appl Polym Sci 9:2251
101. Richters P (1970) Macromolecules 3:262

102. Billingham NC, Fearon P, Whiteman D (1999) The use of chemiluminescence imaging to evaluate stabiliser performance in polymers. In: Proceedings Addcon World 99, Prague, Czech Republic, 27–28 Oct 1999. Smithers Rapra, Shrewsbury, UK
103. European Commision (2002) Commission Directive 2002/72/EEC of 6 August 2002 relating to plastics materials and articles intended to come into contact with foodstuffs, Off J Eur Comm L220, 15.08.2002, pp 18–58
104. Hawkins WL, Worthington MA, Matreyek W (1960) J Appl Polym Sci 3:277
105. Scott G (1995) Biochem Soc Symp 61:235
106. Lawson G, Barkby CT, Lawson C, Fresenius J (1996) Anal Chem 354:483
107. Nagy K, Epacher E, Staniek P, Pukanszky B (2003) Polym Degrad Stab 82:211
108. Villberg K, Veijanen A, Gustafsson I, Wickstrom K (1997) J Chromatogr A 791:213
109. Till DE, Ehntholt DJ, Reid RC, Schwartz PS, Sidman KR, Schwope AD, Whelan RH (1982) Ind Eng Chem Prod Res Dev 21:106
110. Garde JA, Catalá R, Gavara R (1998) J Food Prot 61:1000
111. Garde JA, Catalá R, Gavara R, Hernandez RJ (2001) Food Addit Contam 18:750
112. Berg BE, Hegna DR, Orlien N, Greibrokk T (1992) J High Resolut Chromatogr 15:837
113. Dopico-Garcia MS, Lopez-Vilarino JM, Gonzalez-Rodriguez MV (2003) J Chromatogr A 1018:53
114. Dopico-Garcia MS, Lopez-Vilarino JM, Gonzalez-Rodriguez MV (2005) Talanta 66:1103
115. Haesen G, Le Geoff B, Glaude P (1981) Mikrochim Acta 75:25
116. Burman L, Albertsson AC, Hoglund A (2005) J Chromatogr A 1080:107

Adv Polym Sci (2008) 211: 159–185
DOI 10.1007/12_2008_140

Published online: 15 March 2008

Migration of Monomeric and Polymeric PVC Plasticizers

Minna Hakkarainen

Department of Fibre and Polymer Technology,
School of Chemical Science and Engineering,
Royal Institute of Technology (KTH), Teknikringen 56–58, 10044 Stockholm, Sweden
minna@polymer.kth.se

Abstract This paper summarizes current literature on the migration of monomeric and polymeric PVC plasticizers from medical materials, food packaging, and toys. Especially highlighted is macromolecular engineering as a tool to increase the plasticizing efficiency and migration resistance for polymeric plasticizers. The effect of branching, molecular weight, end-group functionality, and polydispersity on plasticizer performance was evaluated by quantification of low-molecular-weight hydrolysis products, measurements of mechanical properties, weight loss, surface segregation, as well as the preservation of material properties during aging. A more migration-resistant polymeric plasticizer that also better preserved its material properties during aging was obtained by combining a low degree of branching, hydrolysis-protecting end-groups, and higher molecular weight.

Keywords Migration · Phthalate · Plasticizer · Poly(butylene adipate) · Poly(vinyl chloride)

Abbreviations

AA	Adipic acid
ATBC	Acetyl tributyl citrate
BBP	Butylbenzyl phthalate
BT	1,2,4-Butanetriol
DBP	Dibutyl phthalate
DBS	Dibutyl sebacate
DCHP	Dicyclohexyl phthalate
DEHA	Di-(2-ethylhexyl) adipate
DEHP	Di-(2-ethylhexyl) phthalate
DEHS	Di-(2-ethylhexyl) sebacate
DINP	Diisononyl phthalate
DIDP	Diisodecyl phthalate
DMA	Dimethyl ester of adipic acid
DOA	Dioctyladipate
DOP	Dioctylphthalate
DPOP	Diphenyl 2-ethylhexyl phosphate
DSC	Differential scanning calorimetry
ESBO	Epoxidized soybean oil
FTIR	Fourier transform infrared spectroscopy
GC	Gas chromatography
HBP	Hyperbranched polymer
HPLC	High performance liquid chromatography
HS	Headspace
LC	Liquid chromatography
MEHP	Mono-(2-ethylhexyl) phthalate
MS	Mass spectrometry
MVDA	Multivariate data analysis
NMR	Nuclear magnetic resonance
NP	Nonylphenol
PBA	Poly(butylene adipate)
PCL	Polycaprolactone
PCL-PC	Polycaprolactone-carbonate
PEG	Polyethylene glycol
PHBV	Poly(3-hydroxybutyrate-co-3-hydroxyvalerate)
PVC	Poly(vinyl chloride)
SEC	Size exclusion chromatography
SPE	Solid phase extraction
SPME	Solid phase microextraction
TMP	Trimethylol propane

1 Introduction

In terms of consumption, poly(vinyl chloride) (PVC) is one of the most important plastic materials available today, providing unmatched material properties in many applications. Phthalate esters have been used as plasticizers to prepare flexible PVC products since the 1920s [1]. In recent years there have been many studies on the possible negative effects of some of the phthalates on human health and on the environment. In 2005, however, the phthalates still constituted almost 90% of the world's plasticizer market [2]. Because of the low molecular weight and the lack of chemical bonds between phthalates and PVC, they can rather easily be extracted from the polymer matrix and phthalate esters are found in most of the environments, in animals, and in humans [3–6]. Many alternative plasticizers have been developed, but it has been difficult to find alternatives that provide as broad range of properties than phthalate esters. Alternative plasticizers on the market include phosphates, aliphatic dibasic esters, trimellitates, pyromellitates, and polymeric plasticizers [7]. In addition, application of surface coating [8–10], multi-layered tubing [11, 12] or surface modification by radiation cross-linking [13–15] have been successfully used to reduce the migration of low-molecular-weight additives to the environment, but often at the expense of other properties such as flexibility, thermal stability, surface characteristics, and appearance [1].

The demand for new alternative plasticizers, especially for medical applications, toys and food packaging, is still strong and non-toxic polymeric plasticizers are one appealing solution. PVC is miscible with many aliphatic polyesters such as polycaprolactone (PCL) [16–21], poly(3-hydroxybutyrate-co-3-hydroxyvalerate) (PHBV) [22], poly(butylene adipate) (PBA) [23–26] poly(caprolactone-carbonate) (PCL-PC) [27, 28] and poly(caprolactone-ethylene glycol-carbonate) [29]. The miscibility is a result of specific interactions between the carbonyl group in the polyester and the CHCl group in PVC [30–33]. As the molecular weight of the plasticizer increases, there are more entanglements and less chain ends per mass of plasticizer. High molecular weight usually means that the glass transition temperature is raised, the tensile strength is increased, the tensile elongation is reduced, and the material becomes more difficult to process than a material plasticized with a low-molecular-weight plasticizer.

One way to improve the performance of polymeric plasticizers could be to change the macromolecular structure from linear to branched. In the same way that low-molecular-weight compounds induce a greater free volume in the material, the higher chain-end density of a branched polymer could provide more mobility in a blend than a linear polymer of the same molecular weight [34]. Hyperbranched polymers reduce the melt viscosity [35] in polymer blends and improve the processability of the material. Fully miscible blends of hyperbranched and linear polymers can be achieved if strong intermolecular interactions, such as hydrogen bonds, are present between the two polymers [36]. Very few studies have been published on the mixing of branched or highly branched polyesters with PVC. However, a highly branched polyester plasticizer based on di-and mono-carboxylic acids, diols, and triols was patented by Huber et al. already in 1967 [37]. A hyperbranched polycaprolactone [38] and a series of highly branched, patented polyesters [39] similar to hyperbranched polycaprolactone were lately reported by Choi and Kwak, while Lindström and Hakkarainen studied a whole range of poly(butylene adipate) plasticizers from linear to highly branched [40–43]. This review summarizes the literature on migration of monomeric and polymeric plasticizer from PVC medical materials, food packaging, and toys. The possibility of improving migration resistance, material performance, and long-term properties through small changes in the design of polymeric plasticizers is especially highlighted.

2 Migration of Monomeric Plasticizers

When plastic materials are used in medical devices, food packaging, or toys, they come in contact with biological fluids or food. These rather aggressive environments accelerate the release of low-molecular-weight additives from the material. Several studies have been done to study the migration of phthalates or other low-molecular-weight compounds in contact with biological fluids, food, or their simulants. Emission of di-(2-ethylhexyl) phthalate (DEHP) from PVC flooring into air and uptake in dust has also been shown [44] and Wensing et al. recently reviewed the literature on plasticizers detected in the indoor environment [45].

2.1 Migration to Biological Fluids and Intravenous Solutions

Contamination of intravenous solutions from PVC-bags was studied by headspace-gas chromatography-mass spectrometry (HS-GC-MS) and liquid chromatography (LC) after the plastic bags were heated in oven for 5–20 min at 120 °C to simulate the heat sterilization conditions [46]. After the heating, di-(2-ethylhexyl) phthalate, mono-(2-ethylhexyl) phthalate (MEHP), phthalic acid, 2-ethylhexanol, phthalide, benzaldehyde, benzoic acid and phenol were identified in the intravenous sodium chloride solutions. Column-switching LC-MS method was developed to be used as a high-throughput technique for direct analysis of di-(2-ethylhexyl) phthalate released from PVC medical devices to intravenous solutions [47]. A similar method utilizing column-switching LC-MS coupled with on-line extraction was used for direct analysis of DEHP and MEHP in blood products to evaluate the human exposure to these compounds [48]. The levels of MEHP and DEHP in the blood products increased with increasing storage time. In addition, whole-blood products had the highest DEHP levels (up to $83\,\mu g\,mL^{-1}$). A method utilizing cloud point extraction coupled with microwave-assisted back extraction followed by gas chromatography was developed to determine di-(2-ethylhexyl) adipate (DEHA) and acetyl tributyl citrate (ATBC) in aqueous solutions [49]. Grafting of PVC surface with polyethylene glycol (PEG) significantly decreased the migration of plasticizers to organic extractants and into oil medium [50]. Grafting also increased the blood compatibility.

2.2 Migration to Food and Food Simulants

The migration of low-molecular-weight plasticizers from PVC to food or food simulants has been investigated in several studies. Diisodecyl phthalate (DIDP) was identified as a main plasticizer in three out of 14 PVC gasket seals, while epoxidized soybean oil (ESBO) was the principal plasticizer in eight of the gasket seals [51]. In the same study, di-(2-ethylhexyl) phthalate was detected in six food samples at a concentration of 2.5 to 8.7 mg kg^{-1}, which is close to or exceeds the specific migration limit of 3 mg kg^{-1}. The concentration of ESBO exceeded 60 mg kg^{-1} in three food samples. In another study, the migration of ESBO, phthalates, di-(2-ethylhexyl) adipate, di-(2-ethylhexyl) sebacate (DEHS) and acetyl tributyl citrate from PVC gaskets of lids for glass jars into oily foods was investigated [52]. The migration of DEHA from food-grade PVC into hard and soft cheese exceeded the upper limit for DEHA migration proposed by EU (18 mg kg^{-1}) [53]. The amount depended on the contact time, fat and moisture content, and consistency of cheese samples. Migration of DEHA from PVC cling films to olive oil food stimulant showed global migration ranging from 20 to 30 mg dm^{-2} [54]. When cheese was stored inside the same films, the level of DEHA in the cheese was 45 mg kg^{-1} cheese after 2 h at 5 °C and increased to 150 mg after 10 days. Castle et al. have in several studies investigated the migration of di-(2-ethylhexyl) adipate from plasticized PVC films into food [55, 56]. Migration increased with both the length of contact time and temperature of exposure, with the highest levels observed where there was a direct contact between the film and food, and where the latter had a high fat content.

Castle et al. also carried out a survey of plasticizer levels in retail foods wrapped in plasticized films or materials with plasticized coatings [57]. Plasticizers found included dibutyl sebacate, acetyl tributyl citrate, dibutyl phthalate (DBP), dicyclohexyl phthalate (DCHP), butylbenzyl phthalate (BBP), and diphenyl 2-ethylhexyl phosphate (DPOP). Levels of plasticizers found in foods were in the following ranges: ATBC in cheese, 2–8 mg kg^{-1}; DBS in processed cheese and cooked meats, 76–137 mg kg^{-1} and 76–137 mg kg^{-1}; DBP, DCHP, BBP, and DPOP found individually or in combination in confectionery, meat pies, cake, and sandwiches, total levels ranging from 0.5 to 53 mg kg^{-1}. Over 50% of DEHA and 40% of ATBC migrated from PVC films into sweetened sesame paste after 240 h at 25 °C [58]. Migration of dioctyl phthalate (DOP) and dioctyl adipate (DOA) plasticizers from food-grade PVC films into ground-meat with different fat contents was determined by GC at 4 °C and

−20 °C [59]. Under similar conditions, DOA migrated more readily than DOP. Fat content of the food increased the migration rate.

Migration of DEHP from plasticized tubing of commercial milking equipment to milk has also been shown [60]. The amount of DEHP increased from $<5\ \mu g\ kg^{-1}$ in samples obtained by hand milking to $30\ \mu g\ kg^{-1}$ in samples obtained by machine milking and up to $1400\ \mu g\ kg^{-1}$ in retail cream samples. Migration of DEHA from rigid PVC bottles to water under dynamic and static conditions was shown by GC [61]. $300\ ng\ L^{-1}$ of nonylphenol (NP) was found in spring water bottled in PVC bottles [62]. In another study, migration of 4-nonylphenol from PVC packaging films into food and food simulants was investigated by HPLC [63]. The amount of NP migrating into heptane was higher than the amount migrating into water or into 4% acetic acid. The migration of NP increased when the samples were reheated in a microwave oven.

2.2.1 The Effect of Radiation on Migration Rate

γ-radiation corresponding to doses used during "cold pasteurization" did not significantly affect the migration of DOA and ATBC from PVC films into olive oil [64]. The amount of DOA that migrated to olive oil exceeded $300\ mg\ L^{-1}$ after 47 h, while the amount of ATBC was under $1\ mg\ L^{-1}$. In thermally treated samples, however, $19.3\ mg\ L^{-1}$ of ATBC had migrated after 60 min at 80 °C. In another study, the effect of γ-radiation on the migration of DEHA and ATBC into isooctane food stimulant was studied by a direct gas chromatographic method [65]. In accordance with the earlier study, DEHA migrated much more readily than ATBC. No radiation-induced transformation of the plasticizers was seen at doses of 5–25 kGy, however the migrated amount increased with increasing radiation dose and contact time. The same authors also studied the effect of γ-radiation on the migration of DOA from food grade PVC film into chicken meat products [66]. While the ionization radiation at intermediate doses did not affect the migration, higher fat content of the foodstuff drastically increased the migration of DOA plasticizer.

2.3 Migration from Toys and Childcare Articles

Several plastic toys were analyzed for their DEHP content by supercritical fluid extraction followed by GC analysis [67]. DEHP recoveries obtained after super-critical fluid extraction were considerably better than those obtained with soxhlet extraction. The results showed that PVC toys normally contained rather high DEHP concentrations. In another study, 72 toys from 17 countries were analyzed for the composition and 64 of these toys contained PVC [68]. Almost all of the soft toys contained 10–40 wt. % of phthalates, the most common phthalates being diisononyl phthalate (DINP) and DEHP. Shaking increased the amount of DEHP that migrated from the PVC child-use articles into saliva from 25 to 499 $\mu g\,g^{-1}$ film [69]. Migration of phthalates from 62 soft toys into saliva simulant by dynamic extraction showed that 47 of these toys contained plasticized PVC [70].

Two laboratory-based linear horizontal agitation methods for studying migration of phthalates from soft PVC toys were developed [71]. Acceptable repeatability within laboratory and reproducibility between laboratories were demonstrated for the GC-MS and the agitation/extraction procedures developed for the analysis of DINP at six different laboratories. The internal exposure of nursery-school children to DEHP was determined by determining the DEHP metabolites mono(2-ethyl-5-hydroxyhexyl)phthalate, mono(2-ethyl-5-oxo-hexyl)phthalate and mono(2-ethylhexyl)phthalate in the first-morning urine by a multidimensional liquid chromatography tandem mass spectrometry method [72]. When calculated in relation to body weight, the concentrations in children were about twice as high as in adults.

2.4 Health Issues Related to Phthalate Ester Plasticizers

Exudation of plasticizers changes the long-term properties of the material and may pose a threat to animals and humans if the plasticizer is passed on to them [73–77]. Phthalate plasticizers have (in in vitro and toxicological animal tests) been shown to have adverse effects on the liver, the reproductive tract, the kidneys, the lungs, the heart, and on the fetus [73, 74]. Phthalate esters have been identified as irritants and immunogenes of respiratory syndromes [75]. Toxicological data, together with limited human-exposure data, also leads to a concern that bis(2-ethylhexyl) phthalate, one of the most common phthalate plasticizers used in medical devices, might be harmful to human fertility and reproduction. In addition, DEHP is a suspected endocrine disruptor [5, 76]. There are also indications that MEHP, the principal hydrolysis product of DEHP, exhibits genotoxic effects in human mucosal cells and lymphocytes [77]. Exposure levels similar to those causing adverse effects in toxicological tests on animals have been found in certain clinical situations, which raises concerns about exposing fetuses, neonates, infants, children, and chronically ill adults to DEHP [73]. In most of the literature, DINP, another common phthalate plasticizer, has been classed as unlikely to pose a risk to the environment or to consumers [78], but there are indications that DINP may lead to increased allergic responses [79].

In 1999 a ban [80] was issued by the European Commission forbidding the use of six of the most common phthalate plasticizers (DINP, DIDP, DEHP, DBP, BBP, DOP) in toys and child-care articles that are intended to be placed in the mouths of children under the age of three. In July 2005, this ban was converted into a directive [81] that forbids the use of three of these phthalate esters, DEHP, DBP, and BBP, in any toys and child-care articles within the

European Union. In addition, the new directive forbids the use of the other three phthalate esters, DINP, DIDP, and DnOP, in toys and child-care articles that can be placed in the mouth. Prior to the emergency ban in 1999, DEHP was the most widely used phthalate plasticizer in Sweden, but since 2000 it has in many products been replaced by DINP. DEHP is, however, still used in health-care products such as tubes, catheters, blood bags, and haemodialysis devices [82]. The new and more strict legislation regarding the use of phthalate ester in PVC products, however, triggers the need for alternative, well-performing plasticizers for PVC.

3 Migration of Polymeric Plasticizers and Their Degradation Products

There have been many studies on the migration of low-molecular-weight plasticizers from PVC, and a few studies have also addressed the extraction or migration of linear polymeric plasticizers and their degradation products from flexible PVC products [26, 83–85]. Low-molecular-weight polyesters exuded faster from PVC sheets than higher-molecular-weight analogues and acetylation of hydroxyl groups reduced the exudation [86]. Migration of poly(butylene adipate) oligomers in the molecular weight range of 300-1100 to olive oil showed that the smallest oligomers migrated 90-fold more readily than the bulk of the plasticizer [26]. Monomers and oligomers were also found to migrate when commercial PVC tubes plasticized with linear poly(butylene adipate) were aged in water at different temperatures [87]. Castle et al. showed that by replacing part of the di-(2-ethylhexyl) adipate plasticizer with polymeric plasticizer the migration was reduced compared to films plasticized with only DEHA or only polymeric plasticizer [88]. Weight loss and water absorption of PVC/PBA blends were highly dependent on the aging temperature, and pH of the aging medium strongly affected the formation of adipic acid [87]. Partial hydrolysis of poly(butylene adipate) and poly(propylene adipate) plasticizers was observed in contact with simulated gastric conditions [89]. However, hydrolysis led mainly to the formation of oligomers and did not result in significant conversion to monomeric products. PBA was added as the principal plasticizer in four out of 14 PVC gasket seals [51]. The concentration of PBA in eight food samples that had been in contact with these PVC gasket seals was in all cases below 30 $\mathrm{mg\,kg^{-1}}$ for the low-molecular-weight fraction under 1000 $\mathrm{g\,mol^{-1}}$. In vitro hydrolysis of poly(1,2-propylene adipate) by digestive fluid liquids was studied to assess the safety of polymeric plasticizers that could migrate into food [90]. In intestinal fluid solution rapid hydrolysis to oligomers took place. However, practically no adipic acid monomer was formed. Almost no hydrolysis took place under simulated gastric and saliva conditions.

A strong influence of temperature on the migration and hydrolysis of plasticizer was seen when PVC/polycaprolactone-carbonate tubes were aged in water or phosphate buffer at different temperatures [27, 28]. During the study, a solid-phase microextraction (SPME) method was developed to quantitatively determine the amount of 6-hydroxyhexanoic acid in aqueous solutions [27]. Figure 1 shows that only small amount of 6-hydroxyhexanoic acid, the final hydrolysis product of PCL-PC, migrated from the blends during ageing at 37 °C and 70 °C and weight loss was very low (Fig. 2). When the temperature was raised to 70 °C, only a minor increase in the amount of 6-hydroxyhexanoic acid was observed and the weight loss after 98 days was still under 1%. If, however, the temperature was raised to 100 °C it resulted in

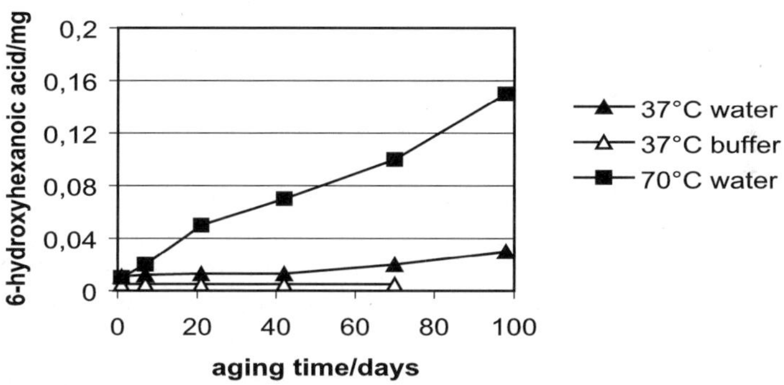

Fig. 1 The amount of 6-hydroxyhexanoic acid that migrated from PVC/PCL-PC to water and buffer solution after different aging times at 37 °C and 70 °C. The amount of 6-hydroxyhexanoic acid is given as mg/g of PVC/PCL-PC blend. Reprinted from [28] with permission of Elsevier. © Elsevier (2003)

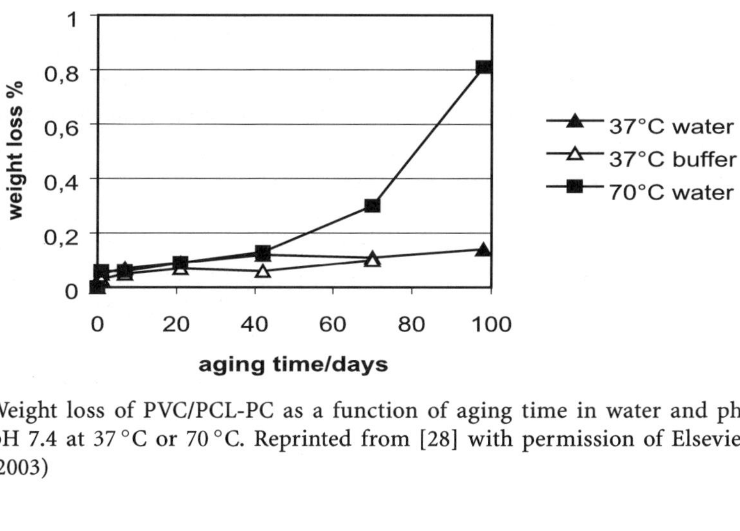

Fig. 2 Weight loss of PVC/PCL-PC as a function of aging time in water and phosphate buffer pH 7.4 at 37 °C or 70 °C. Reprinted from [28] with permission of Elsevier. © Elsevier (2003)

rapid hydrolysis of PCL-PC and lead to almost exhaustive exudation of plasticizer. In addition to 6-hydroxyhexanoic acid, 6-hydroxyhexanoic acid dimer, caprolactone, different carboxylic acids, acetophenone and phenol were identified. Almost no changes in the material or its surface composition were observed during 98 days at 37 °C in water or phosphate buffer. Changes in the fourier transform infrared spectra (FTIR) indicating migration of PCL-PC towards the surface of the PVC/PCL-PC tubing were observed first after 70 days at 70 °C.

The promising results by Huber et al. [37] and Choi and Kwak [38] with a reduction in the migration of highly branched polyesters to hexane and polymer substrates suggested that the introduction of branches in the plasticizer structure could be an effective way to prevent the release of plasticizer also in aqueous environments. Volatility, extractability, and exudation tests for PVC/hyperbranched polycaprolactone showed that there was no plasticizer migration even at very harsh conditions, while 7–78% of additives in PVC/DEHP migrated out of the samples [91].

3.1 Migration of Polymeric Plasticizers During Sterilization

Strips made of different PVC/PCL-PC blends were sterilized in an autoclave at 125 °C for 20 min to study the migration of low-molecular-weight compounds from the blends during sterilization. In addition to PCL-PC, the blends contained different additives such as epoxidized soybean oil. After sterilization, the low-molecular-weight compounds were extracted by SPME. Several SPME methods were tested to ensure the extraction of all low-molecular-weight compounds from water. As a result of these tests, it was decided to extract all the samples with polyacrylate and carbowax/divinylbenzene SPME fibers. GC-MS analysis of the extractions showed that only a few low-molecular-weight compounds migrated from the materials during sterilization, Fig. 3. Some 6-hydroxyhexanoic acid and hexadecanoic acid migrated from all the PVC/PCL-PC samples. Two different synthesis routes were used to prepare the PCL-PC. In the first synthesis, one of the monomers used was diphenyl carbonate, which resulted in the formation and migration of phenol and acetophenone impurities from the material (Fig. 3, chromatogram above). No

Fig. 3 GC-MS chromatograms showing the low-molecular-weight compounds migrating from PVC/PCL-PC during steam sterilization. Some 6-hydroxyhexanoic acid (main product) and hexadecanoic acid migrated from all the PVC/PCL-PC samples. If diphenyl carbonate was used for the synthesis of PCL-PC phenol and acetophenone impurities were also present and released from the material (chromatogram *above*). When another synthetic route was developed, no phenol or acetophenone was detected (chromatogram *below*) ▶

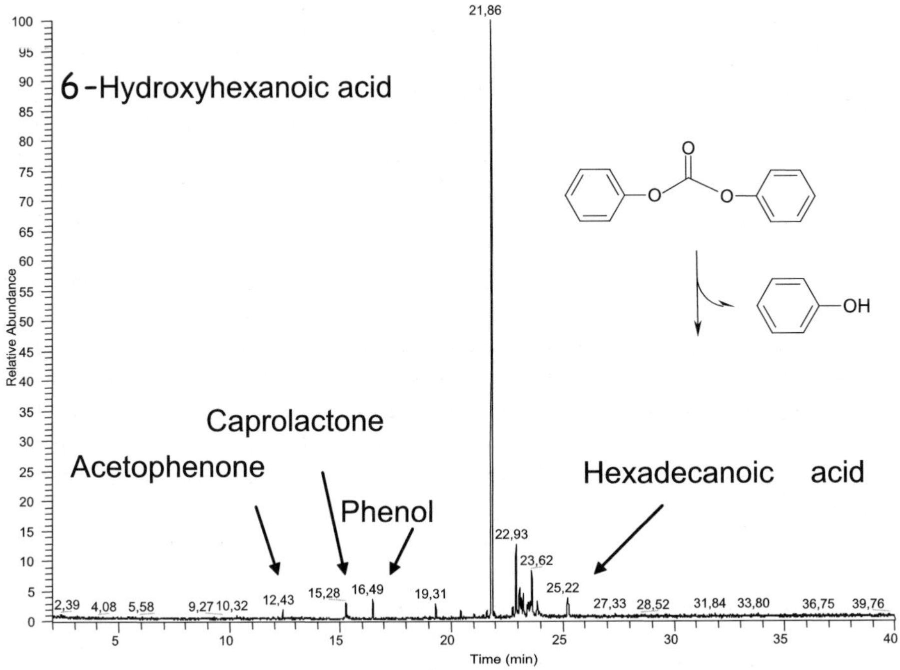
6-Hydroxyhexanoic acid
Caprolactone
Acetophenone
Phenol
Hexadecanoic acid
OH
Relative Abundance
Time (min)

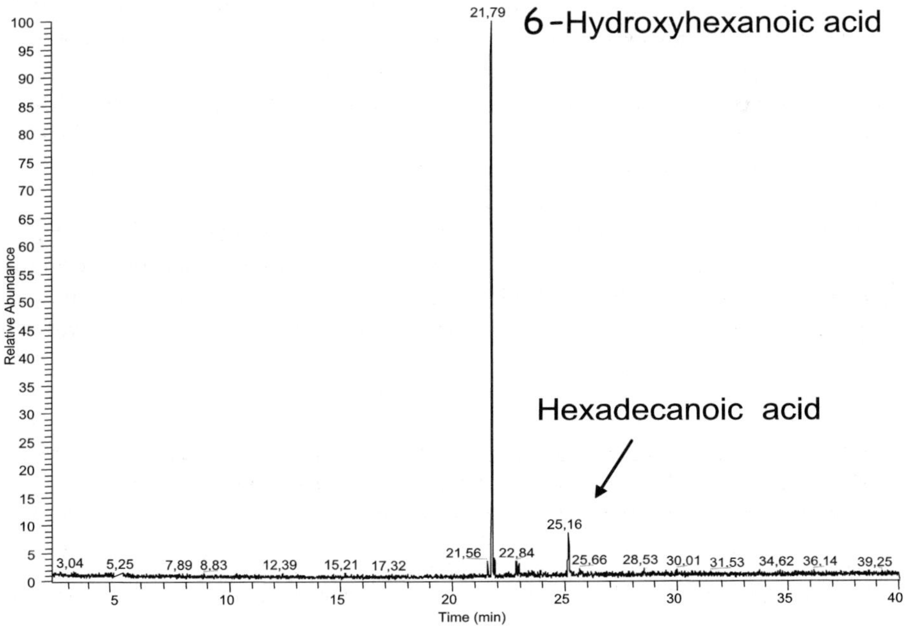
6-Hydroxyhexanoic acid
Hexadecanoic acid
Relative Abundance
Time (min)

phenol or acetophenone was detected when PCL-PC was synthesized without diphenyl carbonate (Fig. 3, chromatogram below).

4 Tuning of Blend Properties and Migration Resistance Through the Architecture of Plasticizer

Lindström and Hakkarainen systematically studied the effect of plasticizer architecture on blend properties and migration resistance [40–43]. Eleven different poly(butylene adipate)s were solution-cast with PVC to elucidate the possibility of improving the miscibility and blend properties by changing the molecular weight, polydispersity, end-group functionality, or degree of branching. The poly(butylene adipate)s were synthesized through polycondensation of 1,4-butanediol and adipic acid or 1,4-butanediol and dimethyl ester of adipic acid. Trimethylol propane (TMP) [40, 41] or 1,2,4-butanetriol (BT) [42, 43] were used as branching agents. The synthesized PBAs were characterized by size-exclusion chromatography (SEC) and nuclear magnetic resonance (NMR), Table 1.

There was a good agreement between the Mark–Houwink α-values obtained from SEC measurements and the TMP content calculated from NMR spectra for polyesters synthesized with TMP as branching agent. The Mark–Houwink α-value is related to the shape of the molecule. Spherical structures have small α-values close to 0 while rod-like structures have values of about 2.0. The low α-values for branched polyesters BT12, BT13, and BT14, i.e., 0.467, 0.414, and 0.305, respectively, indicate a more spherical structure and considerable branching, especially in the case of BT14. The increasing degree of branching for BT12, BT13, and BT14 was also confirmed by ^{13}C-NMR and FTIR [42].

4.1 Development of the Solid-Phase Extraction Method

A solid-phase extraction (SPE) method was developed for the extraction of hydrolysis products from PVC/PBA blends [92, 93]. The expected final hydrolysis products of PBA, 1,4-butanediol and adipic acid, are highly water-soluble and they also have very different pKa values, which makes the simultaneous extraction from water a demanding task. Four different non-polar SPE columns were tested for the extraction: C8 silica, C18 silica, C18 silica (EC) and ENV+ resin. C8 and C18 use non-polar interactions to extract organic compounds from aqueous matrices, and their non-polar characteristics increase with increasing carbon chain length, end-capping (EC) by trimethyl silane is used to reduce the number of silanol groups present at the surface. ENV+ is a highly cross-linked sorbent that has a higher capacity than nor-

Table 1 Average molecular weights, polydispersity index, mole-percent branching agent, branching frequency, Mark–Houwink-α value, and melting temperature for the synthesized polyesters. L = linear PBA, TMP = trimethylol propane was used as a branching agent, BT = 1,2,4-butanetriol was used as branching agent

Polyester	Mn (g/mol)	Mw (g/mol)	PDI	TMP [a] (%)	Mark–Houwink-α	Branching frequency
L1 [b]	4400	7400	1.7			
L2 [b]	8400	14 800	1.8			
TMP3 [b]	1700	4400	2.6	0.6		
L4 [c]	2300	3300	1.4		0.582	
L5 [c]	4100	6100	1.5		0.604	
L6 [b]	7800	14 500	1.9		0.608	
TMP7 [c]	4200	6900	1.6	0.8	0.55	
TMP8 [b]	5000	8000	1.6	1.8	0.543	
TMP9 [c]	5400	8800	1.6	0.4	0.597	
TMP10 [b]	9700	17 500	1.8	0.2	0.629	
L11 [b]	2400	4700	2.0		0.557	0.00
BT12 [b]	2100	5000	2.3		0.467	0.19
BT13 [b]	1050	3350	3.2		0.414	0.54
BT14 [b]	700	6700	9.6		0.305	1.23

[a] Measured by NMR
[b] Synthesized from dimethyl ester of adipic acid, DMA
[c] Synthesized from adipic acid, AA

mal C18 sorbents, and it is especially suited for extracting polar analytes from aqueous matrices. Acidic methanol (0.1% HCl) as elution solvent showed good repeatability and improved GC separation due to methylation of dicarboxylic acids to methyl esters. The extraction efficiencies for the different solid phases were compared by analyzing the relative amounts of the different analytes extracted (sample concentration 300 ng μl^{-1}), Fig. 4. ENV+ was finally chosen as a column material because it had a higher extraction efficiency for 1,4-butanediol than the other solid phases. The SPE extraction procedure by using ENV+ column is shown schematically in Fig. 5.

Known amounts of adipic acid and 1,4-butanediol were dissolved in acidic methanol and analyzed in order to determine the extraction efficiencies. Extraction efficiencies, 79% for adipic acid and 76% for 1,4-butanediol, were calculated by dividing the slope of the calibration curve for extracted samples by the slope of the calibration curve for non-extracted samples. Bearing in mind the different characters of the extracted compounds, the recoveries obtained by this single-step extraction were good. Hirschlag and Köster earlier used SPE to extract hydroxyacids and dicarboxylic acids from aqueous media [94]. Their recovery for adipic acid was 67% using a SDB-1 column from

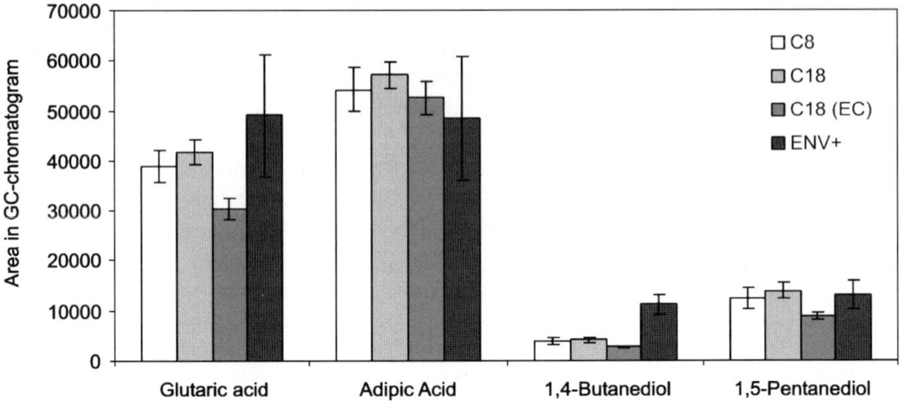

Fig. 4 Comparison of the different solid-phase extraction columns tested for the extraction of adipic acid, 1,4-butanediol, and the internal standards glutaric acid and 1,5-pentanediol from water. Relative intensities were compared for a sample concentration of 300 ng/μl. Reprinted from [92] with permission of Elsevier. © Elsevier (2004)

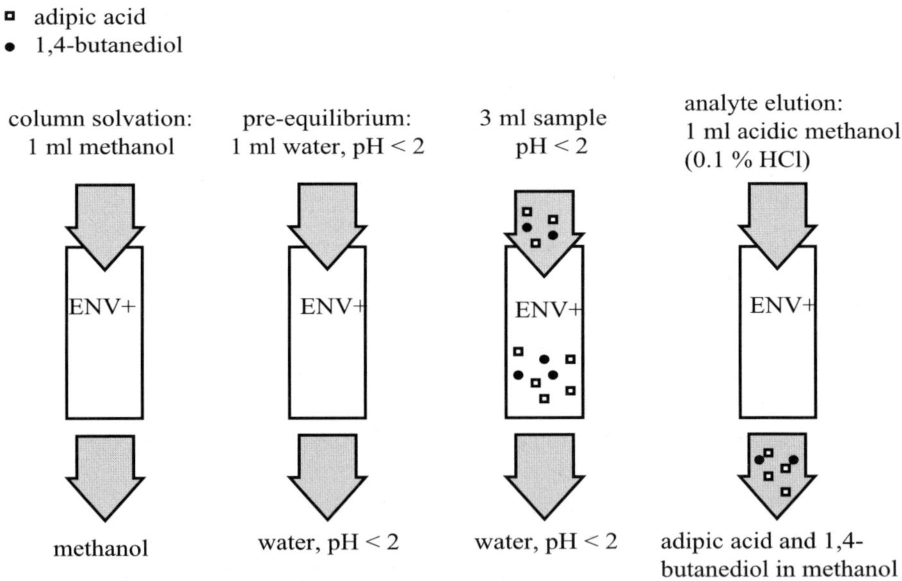

Fig. 5 Schematic presentation of the SPE procedure for the extraction of adipic acid and 1,4-butanediol from water by ENV+ SPE column

J.T. Baker. Relative standard deviations, detection limits, and quantification limits for adipic acid, 1,4-butanediol, and internal standards are presented in Table 2 [92].

Table 2 Detection limits (S/N = 3), quantification limits (S/N = 10), and relative standard deviations (RSD) for adipic acid, 1,4-butanediol, and the internal standards. Reprinted from [92] with permission of Elsevier. © Elsevier (2004)

Analyte	S/N = 3 Concentration (ng/μl)	S/N = 10 Concentration (ng/μl)	RSD (%) [a]
Glutaric acid, dimethyl ester	1	2	
Adipic acid, dimethyl ester	1	3	4
1,4-Butanediol	2	6	3
1,5-Pentanediol	2	7	

[a] RSD values for extracted degradation products, corrected with the respective internal standard, were calculated from six extractions made on the same day.

4.2 Characterization of PVC/Poly(butylene Adipate) Blends

4.2.1 Miscibility

The basis for the PVC/polyester miscibility are the specific interactions involving the carbonyl oxygen in the polyester and the CHCl-group in PVC [19, 30–33]. These interactions can be monitored by a shift in the carbonyl group absorption band in the infrared specta [18, 21, 95]. The shifting of the carbonyl group in the studied PVC/PBA blends compared to the carbonyl group in the spectra of pure polyesters was, thus, interpreted as proof of intermolecular interactions [40, 42]. The miscibility was also confirmed by differential scanning calorimetry (DSC), which showed a single composition-dependent glass transition for all blends except the hyperbranched polyester BT14. In the PVC/BT14 blends, two glass transitions were seen. CH_x/COO ratio of approximately 3–4 has been suggested as the lower limit for miscibility in PVC/polyester blends [19, 23]. Under this ratio, the polyester backbone becomes too rigid, and the free rotation of the chain is hindered, which prevents the formation of intermolecular interactions. On the other hand, at ratios higher than 10–12, the concentration of interaction centers becomes too low. The CH_x/COO ratio for the synthesized PBAs decreased from 4 for the linear PBA to less favorable values below 3 as branching increased.

4.2.2 Morphology and Surface Segregation

Scanning electron microscopy and X-ray elemental analysis were used to examine the surface morphology and composition of the films plasticized with

40 wt. % of polyesters L6, L11, TMP7, TMP8, BT12, and BT13 [42]. The surfaces of films containing polyesters L6, L11, and TMP8 and BT12 showed crystalline domains separated by an amorphous matrix while crystals at the surface of TMP7 were more densely packed. The films plasticized with 40 wt. % BT13 contained much fewer crystals than the other films and more widespread amorphous areas. X-ray elemental analysis of the films indicated a slight enrichment of the polyester at the film surfaces. The ratio of carbonyl groups to chlorine atoms at the surface obtained from the FTIR spectra of the PVC/PBA blends can also be used as a measure of the surface segregation in the material. A certain degree of surface enrichment of one of the components in a blend may also occur in totally miscible blends in order to lower the surface free energy of the system [96]. However, the greater the segmental miscibility, the less surface segregation is usually observed. The films containing 40 and 60 wt. % polymeric plasticizers synthesized from adipic acid (AA), showed much higher values for the carbonyl/chlorine ratio than those synthesized from dimethyl ester of adipic acid (DMA) [40]. This is in agreement with the qualitative surface characterization, where the blends containing polyesters synthesized from adipic acid showed polyester-like surface spectra at a concentration of 40 wt. % plasticizer. In addition to methyl ester end-groups, a low degree of branching reduced the surface segregation.

4.2.3 Mechanical Properties

The mechanical properties of the solution-cast PVC/PBA films were evaluated by tensile testing in order to study the relationship between plasticizer efficiency and plasticizer characteristics such as chemical structure, molecu-

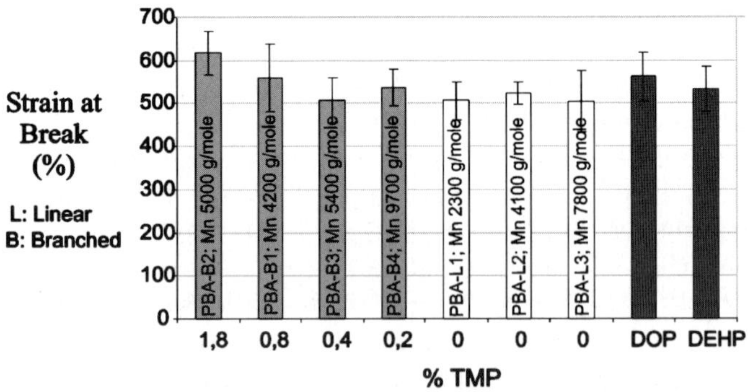

Fig. 6 Strain at break vs. branching agent content of PVC films containing 40 wt. % of linear PBA or branched TMP-based plasticizers. *L* Linear polyester plasticizer, *B* Branched polyester plasticizer. Reprinted from [40] with permission of Wiley-VCH Verlag GmbH & Co. © Wiley-VCH Verlag GmbH & Co. (2006)

lar weight, polydispersity, and degree of branching [40, 42]. Practically all the films showed yield points at concentrations of 20 and 60 wt. % of plasticizer, whereas all the films containing 40 wt. % of plasticizer showed elastomeric stress-strain curves. PVC/TMP8 and PVC/DEHP, also showed elastomeric stress-strain curves at concentrations of 20 and 60 wt. %, respectively. The tensile stress was similar for all the films plasticized with polyester plasticizers. Most desirable mechanical properties were obtained for polyester plasticizers with low degrees of branching. Introducing branches to the structure generates more mobile chain-ends and hence increases the free volume in the material. Increasing the degree of branching however also affects the flexibility of the chains and this may affect the miscibility between the polymeric plasticizer and PVC. Figure 6 shows the strain at break for PVC/PBA blends, where TMP was used as a branching agent.

4.3 Migration from PVC/Poly(butylene Adipate) Blends

4.3.1 Migration of Adipic Acid and 1,4-Butanediol

The PVC/poly(butylene adipate) films were aged in water to evaluate how the polyester design influences the sensitivity towards hydrolysis and the migration resistance [41, 43]. The migrants were extracted from water by SPE and identified and quantified by gas chromatography-mass spectrometry. The films plasticized with linear polyesters L4–6 and branched polyesters TMP7–10, were aged in water at 37 °C for 10 weeks [41] and the films plasticized with DMA-based polyesters L11 and BT12-14 were aged in water at 24 °C for 6 weeks [43]. The amount of monomeric degradation products, i.e., adipic acid and 1,4-butanediol, that migrated from the films was low, but there were significant differences between the films depending on the plasticizer design. In the first series, plasticized with branched polyesters containing TMP, the films plasticized with polyesters having methyl ester end-groups (DMA-based polyesters), i.e., L6, TMP8, and TMP10, were considerably more migration-resistant than the films plasticized with polyesters with carboxyl acid end-groups (AA-based polyesters), i.e., L4, L5, TMP7, and TMP9 [41]. The films plasticized with polyesters with branched structure in combination with methyl ester end-groups had the highest migration resistance.

The films plasticized with AA-based polyesters also showed much higher levels of polyester enrichment at the surface than films plasticized with DMA-based polyesters, which indicates better miscibility between PVC and polyesters containing methyl ester end-groups. Stronger interactions between plasticizer and matrix and less attractive interactions with water make methyl ester end-groups less inclined to be extracted by aqueous media than species with carboxylic acid end-groups. Methyl ester end-groups also have a stabilizing effect towards hydrolysis, which should further reduce the migration of monomeric degradation products from the material. The catalytic effect of carboxylic acid end-groups on the hydrolysis of polyesters and the stabilizing effect of end-capping has been discussed in several studies concerning the hydrolysis of aliphatic polyesters [97–99].

The amount of low-molecular-weight migrants from films containing DMA-based polyesters L11 and BT12–14 aged in water at 24 °C for 6 weeks increased with increasing degree of branching [43]. An increase in the degree of branching in BT polyesters was accompanied by an increase in the hydroxyl end-group content, greater polydispersity, and a reduction in molecular weight, which further facilitates migration to water. The lower molecular weight and slightly higher polydispersity of the linear polyester L11 could explain the greater migration from films containing L11 than from those containing L6, despite the slightly higher aging temperature and longer aging time of the latter. In agreement with the mechanical properties, desirable migration behavior seems to results from introduction of polyester plasticizers with low degree of branching.

4.3.2 Mass Loss and Water Absorption

The stability of PVC/PBA blends in water was also followed by measuring weight loss, water absorption, and pH-changes in the aging media [41, 43]. The weight losses were generally low for all the films, and in good agreement with the amount of adipic acid and 1,4-butanediol detected, Fig. 7. The major part of the weight loss that was observed for PVC/PBA films after aging at 37 °C was due to the release of oligomeric species [41]. The difference between the amount of monomeric products formed and the weight loss of the films plasticized with AA-based polyesters (L4, L5, TMP7, and TMP9), indicated that the monomeric species corresponded to 38–54% of the overall weight loss. For films plasticized with DMA-based polyesters (L6, TMP8, and TMP10), however, the monomeric species amounted to only 2–12% of the total weight loss. This was explained by the hydrolysis-protecting function of the methyl ester end-groups. Films containing the linear DMA-based polyester L6 showed a weight loss similar to that of films containing linear AA-based polyesters. In agreement with GC-MS results, the plasticizer with best migration resistance was obtained by combining a slightly branched polymer structure and hydrolysis-protecting methyl ester end-groups.

For films plasticized with linear polyester L11 and branched polyesters BT12-14, the relative weight loss increased with increasing degree of branching [43]. This weight loss could be explained as being due to a greater mobility in the more open structure resulting from a more branched structure, together with the lower molecular weight and higher polydispersity of the more branched polyesters. Oligomeric species formed from especially the hy-

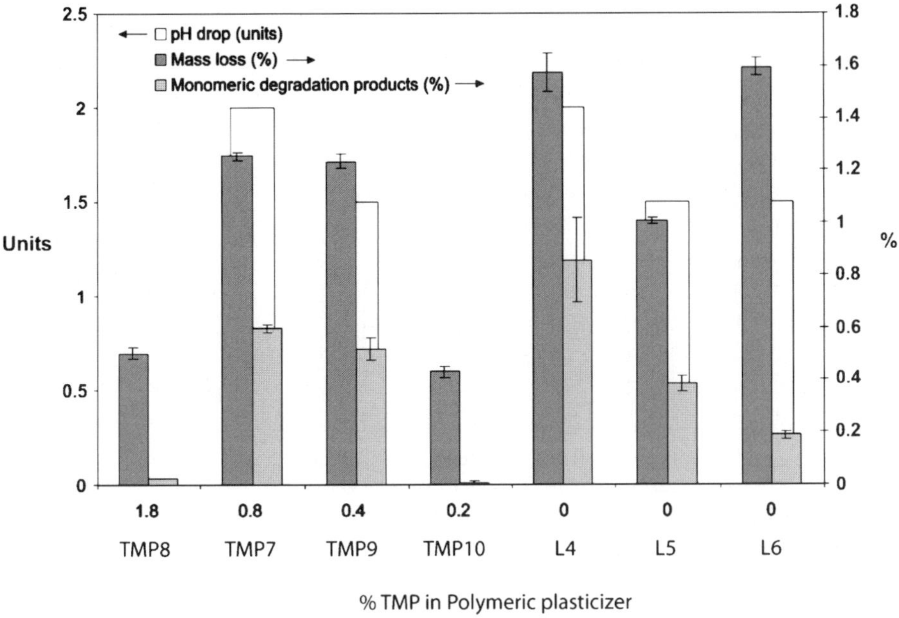

Fig. 7 Weight percent of monomeric degradation products migrating from PVC blends with linear PBA or TMP-based branched PBA after aging in water at 37 °C for 10 weeks, compared to mass loss and pH-changes of the aging medium. The amount of monomeric degradation products is given as amount/theoretical amount if all the polyester is hydrolyzed to monomeric products. Reprinted from [41] with permission of Wiley-VCH Verlag GmbH & Co. © Wiley-VCH Verlag GmbH & Co. (2007)

perbranched polyester BT14 are also more water soluble, and hence more inclined to migrate, as a consequence of the higher concentration of hydroxyl groups in the structure observed by FTIR. However, only a small amount of adipic acid was released from films containing the highly branched BT14. This is probably due to fewer methyl ester end-groups and less residual DMA in BT14 compared to in L11, BT12 or BT13.

The relative weight loss was greater for films containing 20 wt. % of polyester than for films with 40 wt. % of polyester. Stronger intermolecular interactions and lower surface enrichment in films with 40 wt. % plasticizer compared to the films with 20 wt. % polyester as shown by FTIR probably decreases the mobility of oligomeric species and reduce their migration rate [42, 43]. Enrichment of the polyester at the surface increases the hydrophilicity of the surface compared with that of films with a higher concentration of PVC at the surface. Hydrophilic surfaces also facilitate the migration of water into the material, and greater water absorption is, thus, expected in a more hydrophilic material. For films plasticized with 20 wt. % of L11 or BT12–14 polyester, a higher concentration of polyester was detected at the surface as the degree of branching increased and the correlation to a greater

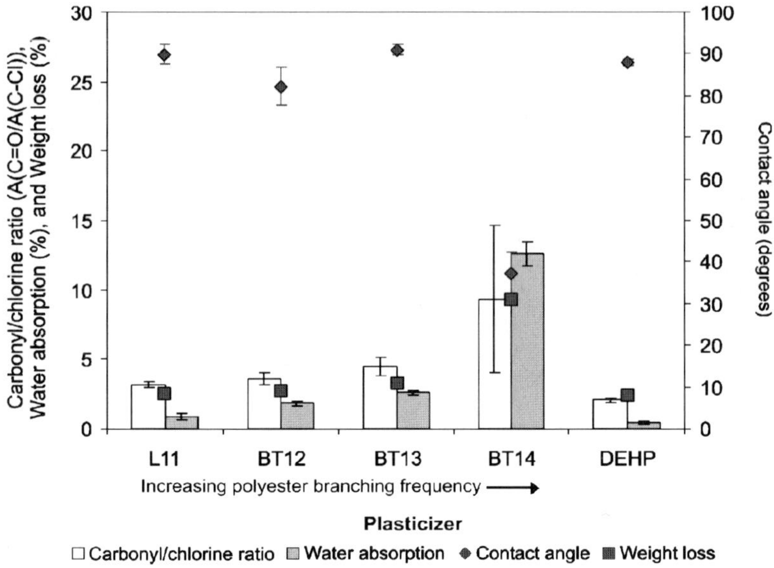

Fig. 8 Water absorption, weight loss, contact angle, and carbonyl/chlorine ratio at the surface of films plasticized with 20 wt. % linear PBA- or BT-based branched plasticizers. Reprinted from [43] with permission of American Chemical Society. © American Chemical Society (2007)

water absorption and a greater weight loss is shown in Fig. 8. A significantly lower contact angle was detected for materials containing the highly branched BT14, which confirms the higher surface hydrophilicity of PVC/ BT14 blends.

In correlation with the similar carbonyl/chlorine ratios and the similar weight losses for all films plasticized with 40 wt. % of L11 or BT12-13, these films absorbed similar amounts of water and showed a similar surface hydrophilicity regardless of the polyester structure, Fig. 9 [43]. At the same time, as stronger intermolecular interactions and a closer packing of the polymer chains lower the release rate of degradation products, they also make it more difficult for small molecules to penetrate the material. This further supports the better miscibility of the 40 wt. % blends, and could explain why similar or lower water absorption was detected in 40 wt. % blends compared to 20 wt. % blends, despite the higher hydrophilicity and higher carbonyl/chlorine ratio of films with 40 wt. % of polyester.

4.4
Multivariate Data Analysis

Multivariate data analysis (MVDA) is a useful tool to evaluate multiple variables at the same time and to describe the relationship between different material properties. Principal component analysis provides an overview of the

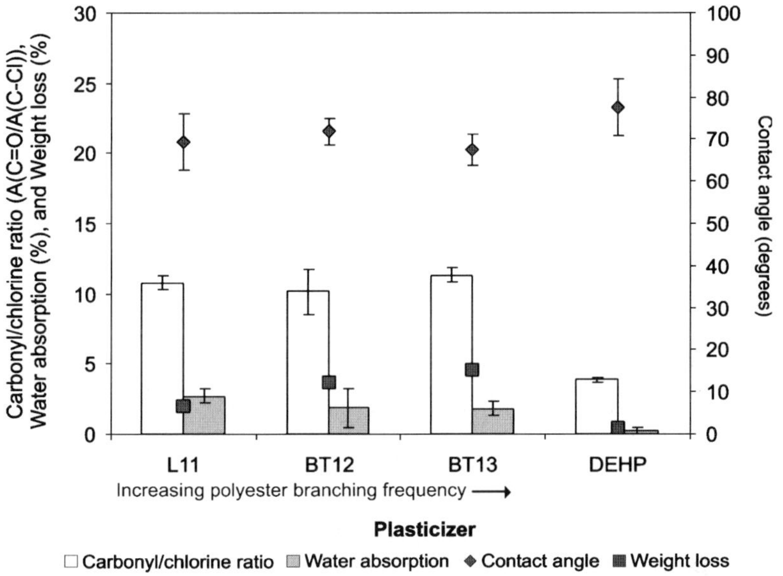

Fig. 9 Water absorption, weight loss, contact angle, and carbonyl/chlorine ratio at the surface of films plasticized with 40 wt. % linear PBA or BT-based branched plasticizers. Reprinted from [43] with permission of American Chemical Society. © American Chemical Society (2007)

data by plotting them in a multidimensional X-space, where the principal components are orthogonal vectors along the directions of variation in the data. Score plots show the relationship among observations (samples), while loading plots show the relationship among variables (properties). MVDA was used to evaluate the different results obtained by tensile testing, FTIR, DSC, migration observations, and material characteristics of films plasticized with polyesters L4–6 or TMP7–10 [41]. Only blends containing 40 wt. % of plasticizer were included in the multivariate data analysis due to the more desirable properties for this blend composition. The loading plot containing the relationship among properties showed that two distinct groups were formed, Fig. 10. One of the groups included the properties DMA, strain, miscibility and surface character, while the other group contained crystallinity, migration, mass loss, and water absorption. Properties within each group are positively correlated, but the two groups are negatively correlated. This means that if the values in one of the groups increase, the values in the other group decrease.

As in the plot of properties, the different films also formed two groups in the score plot, Fig. 11. Films containing TMP8, TMP10, DOP, and DEHP possessed similar properties as they are all placed in the upper right-hand quadrant of the plot, while films containing L4, L5, L6, TMP7, and TMP9 placed in the lower left-hand quadrant displayed opposite properties. No strong or moderate outliers were found among the samples.

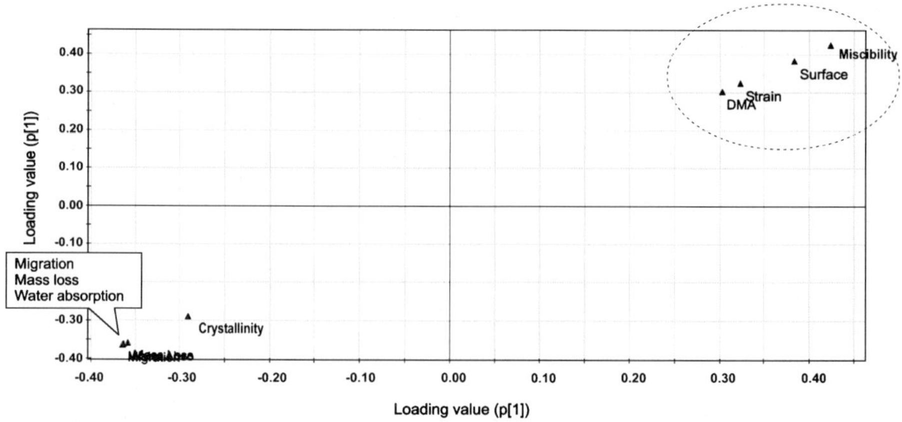

Fig. 10 Loading scatter plot including mechanical and material properties and migration behavior of PVC/PBA films containing 40 wt % of linear PBA or branched TMP-based plasticizers. *Dotted ellipse* properties that are positively correlated to observations TMP8, TMP10, DOP, and DEHP (also marked by a *dotted ellipse*) in Fig. 10. Properties placed in the opposite quadrant of the plot are negatively correlated to the variables inside the dotted ellipse. Reprinted from [41] Wiley-VCH Verlag GmbH & Co. © Wiley-VCH Verlag GmbH & Co. (2006)

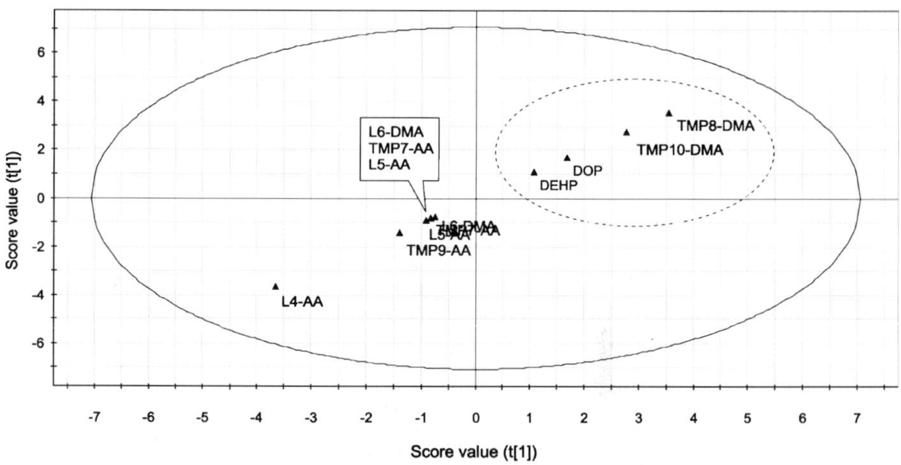

Fig. 11 Score scatter plot including the different samples of PVC/PBA films containing 40 wt. % of plasticizer. *Continuous ellipse* Hotelling's $T^2(0.05)$, *dashed ellipse* observations with favorable properties. Reprinted from [41] with permission of Wiley-VCH Verlag GmbH & Co. © Wiley-VCH Verlag GmbH & Co. (2006)

Films containing plasticizers TMP8, TMP10, DOP, and DEHP (marked by a dotted ellipse, Fig. 11) have high values for DMA, strain, miscibility and surface character. At the same time, they show low values for crystallinity, migration, mass loss, and water absorption, which is a highly attractive of combination of properties for a flexible film material. Blends of PVC with TMP8 and TMP10 have higher scores (i.e., a placement further from the origin in Fig. 11) than blends with DOP and DEHP. This indicates even more desirable properties for PVC films plasticized with polyesters TMP8 and TMP10 compared to films plasticized with traditional phthalate plasticizers. The principal component analysis clearly shows that blends plasticized with TMP8 and TMP10 stand out from the rest of the PVC/PBA blends included in the analysis. The migration of monomeric degradation products and the weight loss were very low for these samples, and the mechanical and material properties were also superior compared to other studied PVC/PBA blends, with values similar to films plasticized with DOP and DEHP. The combined effect of hydrolysis-protecting methyl ester end-groups, a molecular weight that is not too low, and a slightly branched structure improved the miscibility, mechanical properties, and migration resistance.

4.5 Preservation of Material Properties During Aging

PVC films plasticized with polyesters L11 and BT12-14 were also studied by FTIR, DSC, and tensile testing both before and after aging in water at 24 °C for 6 weeks to evaluate the preservation of material properties for the different

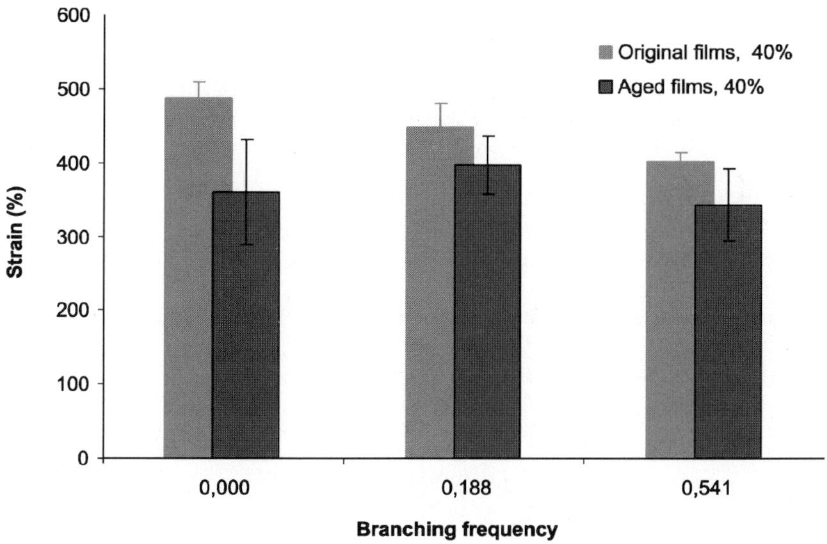

Fig. 12 Strain at break before and after aging in water at 24 °C for 6 weeks for PVC films plasticized with linear or BT-based branched PBA plasticizers at a concentration of 40 wt. % Reprinted from [43] with permission of American Chemical Society. © American Chemical Society (2007)

blends during aging [43]. The films containing 40 wt. % of polyester plasticizers, L11, BT12, or BT13 all showed a significant shift of the carbonyl absorption band in their FTIR spectra before aging. After aging, only the blends containing branched polyesters BT12 and BT13 still showed a significant shift in the carbonyl absorption band. The preservation of desirable material properties during aging was, thus, favored when branched polyester was used as polymeric plasticizer. Films containing DEHP also showed a reduction in the shift of the absorption band as a result of aging. The preservation of mechanical properties was evaluated through tensile testing of films before and after aging in water for 6 weeks. Among the original films, samples plasticized with linear polyester L11 showed the longest elongation at break. However, Fig. 12 shows that during aging, the films plasticized with the slightly branched polyester BT12 proved to be the most ductile. The use of polymeric plasticizers with low degree of branching, thus, helps to preserve good mechanical properties during aging. This is in good agreement with the results from the TMP-series where the slightly branched TMP8 polyester increased the ductility of solution-cast PVC/PBA films compared to films plasticized with the linear polyesters.

References

1. Rahman M, Brazel CS (2004) Prog Polym Sci 29:1223
2. Tullo AH (2005) Chem Eng News 83:29
3. Jaeger RJ, Rubin RJ (1972) New Engl J Med 287:1114
4. Jaeger RJ, Rubin RJ (1970) Science 170:460
5. Latini G, De Felice C, Verrotti A (2004) Reprod Toxicol 19:27
6. Koch HM, Drexler H, Angerer J (2003) Int J Hyg Environ Health 206:77
7. Krauskopf LG (2003) J Vinyl Addit Technol 9:159
8. Messori M, Toselli M, Pilati F, Fabbri E, Fabbri P, Pasquali L, Nannarone S (2004) Polymer 45:805
9. Lee JH, Kim KO, Ju Ym (1999) J Biomed Mater Res 48:328
10. Lakshmi S, Jayakrishnan A (1998) Artif Organs 22:222
11. Bourdeaux D, Sautou-Miranda V, Bagel-Boithias S, Boyer A, Chopineau J (2004) J Pharmaceut Biomed 35:57
12. Di Landro L, Capone C, Inzoli F, Malacari PE (2005) J Vinyl Addit Technol 11:111
13. Jayakrishnan A, Sunny MC (1996) Polymer 37:5213
14. Jayakrishnan A, Lakshmi S (1998) Nature 396:638
15. Ito R, Seshimo F, Haishima Y, Hasegawa C, Isama K, Yagami T, Nakahashi K, Yamazaki H, Inoue K, Yoshimura Y, Saito K, Tsuchiya T, Nakazawa H (2005) Int J Pharm 303:104
16. Hubbell DS, Cooper SL (1977) J Appl Polym Sci 21:3035
17. Chiu F-C, Min K (2000) Polym Int 49:223
18. Coleman MM, Zarian J (1979) J Polym Sci Polym Phys 17:837
19. Prud'homme RE (1982) Polym Eng Sci 22:90
20. Clark MBJ, Burkhardt CA, Gardella JAJ (1991) Macromolecules 24:799
21. Karal O, Hamurcu EE, Baysal BM (1997) Polymer 38:6071
22. Choe S, Cha Y-J, Lee H-S, Yoon JS, Choi H (1995) J Polym 36:4977

23. Ziska JJ, Barlow JW, Paul DR (1981) Polymer 22:918
24. Oriol-Hemmerlin C, Pham QT (2000) Polymer 41:4401
25. Lee S, Lee JG, Lee H, Mumby SJ (1999) Polymer 40:5137
26. Castle L, Mercer AJ, Gilbert J (1991) Food Addit Contam 8:565
27. Hakkarainen M (2003) J Chromatogr A 1010:9
28. Hakkarainen M (2003) Polym Degrad Stab 80:451
29. Ferruti P, Mancin I, Ranucci E, De Felice C, Latini G, Laus M (2003) Biomacromolecules 4:181
30. Garton A (1983) Polym Eng Sci 23:663
31. Garton A, Aubin M, Prud'homme RE (1983) J Polym Sci Polym Lett 21:45
32. Garton A, Cousin P, Prud'homme RE (1983) J Polym Sci Polym Phys 21:2275
33. Varnell DF, Moskala EJ, Painter PC, Coleman MM (1983) Polym Eng Sci 23:658
34. Han YK, Um JW, Im SS, Kim BC (2001) J Polym Sci Polym Chem 39:2143
35. Kim YH, Webster OW (1992) Macromolecules 25:5561
36. Massa DJ, Shriner KA, Turner SR, Voit BI (1995) Macromolecules 28:3214
37. Huber CF, Foulks HCJ, Aylesworth RD (1967) US Patent 3 331 802
38. Choi J, Kwak S-Y (2004) PMSE Preprints 90:449
39. Kwak S-Y, Choi J-S (2004) US Patent 6 747 092
40. Lindström A, Hakkarainen M (2006) J Appl Polym Sci 100:2180
41. Lindström A, Hakkarainen M (2007) J Appl Polym Sci 104:2458
42. Lindström A, Hakkarainen M (2007) J Polym Sci Polym Phys 45:1552
43. Lindström A, Hakkarainen M (2007) Biomacromolecules 8:1187
44. Clausen PA, Hansen V, Gunnarsen L, Afshari A, Wolkoff P (2004) Environ Sci Technol 38:2531
45. Wensing M, Uhde E, Salthammer T (2005) Sci Total Environ 339:19
46. Arbin A, Jacobsson S, Hänninen K, Hagman A, Östelius J (1986) Int J Pharmac 28:211
47. Inoue K, Higuchi T, Okada F, Iguchi H, Yoshimura Y, Sato A, Nakazawa H (2003) J Pharmacol Biomed Anal 31:1145
48. Inoue K, Kawaguchi M, Yamanaka R, Higuchi T, Ito R, Saito K, Nakazawa H (2005) Clin Chim Acta 385:159
49. Zygoura PD, Paleologos EK, Riganakos KA, Kontominas MG (2005) J Chromatogr A 1093:29
50. Lakshmi S, Jayakrishnan A (1998) Artif Org 22:222
51. Ezerskis Z, Morkunas V, Suman M, Simoneau C (2007) Anal Chim Acta 604:29
52. Frankhauser-Noti A, Grob K (2006) Trends Food Sci Technol 17:105
53. Goulas AE, Anifantaki KI, Kolioulis DG, Kontominas MG (2000) J Dairy Sci 83:1712
54. Petersen JH, Naamansen ET, Nielsen PA (1995) Food Addit Contam 12:245
55. Startin JR, Sharman M, Rose MD, Parker I, Mercer AJ, Castle L, Gilbert J (1987) Food Addit Contam 4:385-398
56. Castle L, Mercer AJ, Startin JR, Gilbert J (1987) Food Addit Contam 4:399
57. Castle L, Mercer AJ, Startin JR, Gilbert J (1988) Food Addit Contam 5:9
58. Goulas AE, Zygoura P, Karatapanis A, Georgantelis D, Konominas MG (2007) Food Chem Toxicol 45:585
59. Kondyli E, Panagiotis Demertzis G, Kontominas MG (1992) Food Chem 45:163
60. Castle L, Gilbert J, Eklund T (1990) Food Contam Addit 7:591
61. Choudhry MS, Lox F, Buekens A, Decroly P (1998) Packag Technol Sci 11:275
62. Loyo-Rosales JE, Rosales-Rivera GC, Lynch AM, Rice CP, Torrents A (2004) J Agric Food Chem 52:2016
63. Inoue K, Kondo S, Yoshie Y, Kato K, Yoshimura Y, Horie M, Nakazawa H (2001) Food Addit Contam 18:157

64. Goulas AE, Kokkinos A, Kontominas MG (1995) Z Lebensm Unters Forsch 201:74
65. Zygoura PD, Goulas AE, Riganakos KA, Kontominas MG (2007) J Food Eng 78:870
66. Goulas AE, Kontominas MG (1996) Z Lebensm Unters Forsch 202:250
67. Marín ML, López J, Sánchez A, Vilaplana J, Jiménez A (1998) Bull Environ Contam Toxicol 60:68
68. Stringer R, Labunska I, Santillo D, Johnston P, Siddorn J, Stephenson A (2000) Environ Sci Pollut Res 7:27
69. Steiner I, Scharf L, Fiala F, Washuttl J (1998) Food Addit Contam 15:812
70. Bouma K, Schakel DJ (2002) Food Addit Contam 19:602
71. Earls AO, Axford IP, Braybrook JH (2003) J Chromatogr A 983:237
72. Koch HM, Drexler H, Angerer J (2004) Int J Hyg Environ Health 207:15
73. Ticker JA, Schettler T, Guidotti T, McCally M, Rossi M (2001) Am J Ind Med 39:100
74. Williams DF, de Jong W, Thomsen M, Rodriguez Farre E, Gatti Am, Paunio MK, Hensten Petersen A, Marquardt H (2002) European Commission Health & Consumer protection directorate-general. Opinion on Medical Devices Containing DEHP Plasticised PVC; Neonates and Other Groups Possibly at Risk from DEHP toxicity. Scientific Committee on Medical Products and Medical Devices of the European Committee. Doc.Sanco/Scmpmd/2002/0010:Final.2002
75. Hill SS, Shaw BR, Wu AHB (2001) Clin Chim Acta 304:1
76. Latini G, De Felice C, Presta G, Del Vecchio A, Paris I, Ruggieri F, Mazzeo P (2003) Environ Health Persp 111:1783
77. Kleinsasser NH, Harreus UA, Kastenbauer ER, Wallner BC, Sassen AW, Staudenmaier R, Rettenmeier AW (2004) Toxicol Lett 148:83
78. French Ministry of the Environment, French Ministry of Health and F. M. o. Work (2003) Risk Assessment of 1,2-Benzenedicarboxylic Acid, Di-C8-10-Branched Alkyl Esters C9-Rich, and Di-Isononyl Phthalate (DINP). European Chemicals Bureau, Ispra, Italy
79. Lee MH, Park J, Chung SW, Kang BY, Kim SH, Kim TS (2004) Int Arch Allergy Imm 134:213
80. Byrne D (1999) The Commission of the European Communities, Commission Decision 1999/815/EC, Official Journal of the European Communities, L315/46
81. The European Parliament and the European Council (2005) Directive 2005/84/EC of the European Parliament and the Council. Official J Eur Communities L344:40
82. Vesterberg A, Hedenmark M, Vass A-M (2005) PVC in Medical Devices. An Inventory of PVC and Phthalates Containing Devices Used in Health Care; Department of the Environment and Sustainability. Karolinska University Hospital, Solna, Stockholm
83. Castle L, Nichol J, Gilbert J (1993) Food Addit Contam 10:523
84. Audic JL, Reyx D, Brosse JC (2003) J Appl Polym Sci 89:1291
85. Thomas NL (2004) J Appl Polym Sci 94:2022
86. Rushton BM, Salomons NS (1969) J Appl Polym Sci 13:2341
87. Wang Q, Storm BK (2005) Macromol Symp 225:191
88. Castle L, Mercer AJ, Gilbert J (1988) Food Addit Contam 5:277
89. Castle L, Nichol J, Gilbert J (1993) Food Addit Contam 10:523
90. Hamdani M, Thil L, Gans G, Feigenbaum AE (2002) J Appl Polym Sci 83:956
91. Choi J, Kwak S-Y (2007) Environ Sci Technol 41:3763
92. Lindström A, Albertsson A-C, Hakkarainen M (2004) J Chromatogr A 1022:171
93. Lindström A, Albertsson A-C, Hakkarainen M (2004) Polym Degrad Stab 83:487
94. Hirschlag H, Koster R (1998) Fresenius J Anal Chem 362:274
95. Canavate J, Pages P, Rodriguez MJ, Perez-Folch J, Carrasco F, Colom X (2001) J Polym Mater 18:33

96. Forrey C, Koberstein JT, Pan DH (2003) Interface Sci 11:211
97. Brown DW, Lowry RE, Smith LE (1980) Macromolecules 13:248
98. Huffman KR, Casey DJ (1985) J Polym Sci Polym Chem 23:1939
99. Pitt CG, Gu ZW (1987) J Control Release 4:283

Subject Index

Printing: Krips bv, Meppel, The Netherlands
Binding: Stürtz, Würzburg, Germany